How to Write a Better Thesis

David Evans[†] • Paul Gruba • Justin Zobel

How to Write a Better Thesis

 Springer

David Evans†
University of Melbourne
Parkville
Victoria
Australia

Justin Zobel
Computing and Information Systems
University of Melbourne
Parkville
Victoria
Australia

Paul Gruba
School of Languages and Linguistics
University of Melbourne
Parkville
Victoria
Australia

ISBN 978-3-319-04285-5   ISBN 978-3-319-04286-2 (eBook)
DOI 10.1007/978-3-319-04286-2
Springer Cham Heidelberg New York Dordrecht London

Library of Congress Control Number: 2014931845

Printed on acid-free paper

Springer is part of Springer Science+Business Media (www.springer.com)

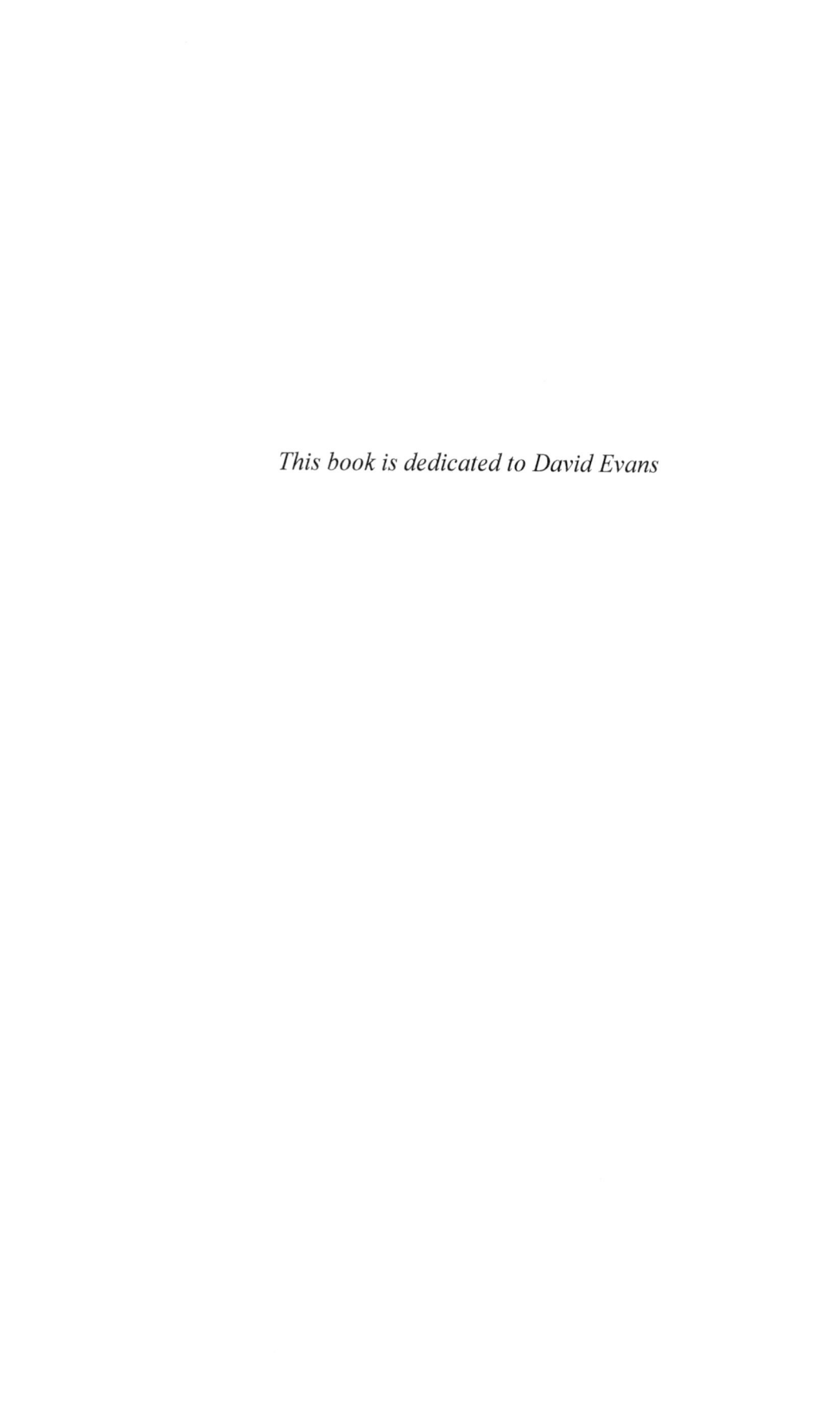

This book is dedicated to David Evans

Preface to the Third Edition

When I began to help to write the second edition with David, my own thesis was still under examination. I had used the first edition of his book, and—perhaps with a bit of bravado—asked David if he would like some assistance when he produced a second edition. He agreed to collaborate. At that time, many of my insights into writing a thesis were based on fresh, personal experience. Sadly, since then, David has passed on. I myself have been lucky enough to gain a full-time academic position and have now supervised several students. More than ever, I can see how important it is to manage the writing process throughout a research project.

I am fortunate to be working with Justin. Not only is he an accomplished supervisor and researcher in his own area of computer science, but he is also the author of a book on writing that is a strong seller in the field. His skills and interests are complementary to mine. Justin works and supervises in science and engineering; I tend to work on qualitative studies in the social sciences.

We have made numerous changes to the second edition. As well as a thorough revision of the text, we have added several new sections that clarify the process of thesis writing. We have eliminated dated advice on word processing and use of computers, for example, and brought forward and updated material concerning written expression. We put greater emphasis on the challenges of thesis writing, the experience of being a research student, the thinking that underlies methods, results, and analysis, and the issues of working with supervisors. Much of the material in this edition is completely new or rewritten, and our book is longer.

Over the years, as I have taught thesis writing seminars, I have used examples of work from my own students to illustrate good writing; I have also used work from John McDonald to show the characteristics of both good and bad theses based on his analysis of examiners' reports. I would like to thank my students, and John, for allowing us permission to use their work here.

For ease of reading, we decided to blend each of our perspectives and experiences—David's, Justin's and my own—into a single collective voice. I hope that you find our collaborative efforts help you to write a better thesis.

Melbourne, February 2011   Paul Gruba

Many years ago I was given a copy of Peter Medawar's *Advice to a Young Scientist*. Though written from the perspective of a biologist, I felt it had lessons for me (in computer science) despite the gulf in research practice between our disciplines. It touched on themes that I felt were lacking in other books on doing research, in particular, what it *felt like* to be a scientist, how one might change and grow as a consequence of doing research, how one might *become* a researcher. It was not that the whole book was on these topics—such a book would probably be rather dull— but I was struck by the perspective that it offered, and how it made Medawar's book different from any number of 'here is a formula for your dissertation' books that tried to reduce being a student to a mechanical process that somehow entirely sidestepped the core of the question of what doing research involves.

Some years ago I was introduced to the second edition of Evans and Gruba's *How to Write a Better Thesis*, and found in it some of those qualities that I had admired in Medawar. It became one of the three or four books I asked every student to read. In working with Paul to produce this new edition, I think we have found ways of strengthening its core messages and have built a text that complements and extends the many 'dissertation' books already on the shelves. Of course, in producing a book like this, it helps enormously to have as a basis a strong existing text, and thus I am grateful to David (who, sadly, I did not have an opportunity to meet) for having created *How to Write a Better Thesis*, and to Paul and David for the revision that produced the second edition.

The framework of this book is the mechanics of thesis writing, but the aim throughout is to help students understand how to conceptualize and approach the problems of producing a thesis, as well as to walk through the details of what a thesis should (or shouldn't) look like. Writing a book like this is something of a journey. It has furthered my understanding of how a student learns to become a researcher, and I have had to sharpen my thinking across a range of topics; it has been illuminating to capture some of the specific lessons learnt from the successes and failures of our students. I hope the book is also a journey for our readers.

A note on style: as Paul has said, we've made no attempt to distinguish between our experiences, including those of David, and have written in the first person. Every example is based on our experience of individual research students, and some of them have been fictionalized to an extent, both to avoid embarrassing people and, in many cases, to make the research more accessible to a general reader. Perhaps confusingly, we've sometimes changed the fictions for the students who were discussed in the previous editions. (Think of it as artistic licence.) In cases where we have quoted from a student's work as an illustration of good work, a full citation is given.

This book rests on our experiences with supervision and advising of upwards of a hundred students, as well as the hundreds of students who have been in our research methods subjects over the past two decades; far too many to name and thank individually, but I am grateful to them for the insights they've brought me and for our experiences together. It is not always obvious to a student how much the supervisor is learning from them, so let this book stand in part as a testament to how mutual a process graduate study can be.

Melbourne, February 2011   Justin Zobel

Introduction

Thesis writing can be challenging for students and supervisors, but one of the many rewards for both parties is to receive positive examiners' reports. I was there when Brian found out that his PhD thesis required just a few minor corrections. He was clearly relieved after years of hard work to discover he had passed with little fuss, but he shouldn't have been too surprised. Brian had written a thesis that, from the start, was well-motivated and purposeful; it was well situated in the field and fluent in the current debates in the discipline; was based on sound principles for data collection; presented results that made it clear what he had achieved; and concluded with his own insightful contributions to the field and observations on how others could pursue further research in the area.

From the start, Brian knew that he had a straightforward task: to convince the examiners that his work had merit, that his data collection and analysis was sound, and that his recommendations were based on firm evidence. In practice, of course, he encountered challenges and worked hard to convey his thinking. Few people have the gift of getting it all down with ease, or with polish. Most students need guidance and editing and criticism, and many struggle during their early attempts to construct and sustain a coherent academic argument. The purpose of this book is to help you to produce a thesis that passes examination.

From the start, good students tend to be independent, confident, and are in the habit of *thinking like a researcher*. Some students have such skills at the beginning, but most have to learn them, and do so by working with their supervisors and other students. In this book, I provide examples of what successful students have done as they have made progress in their work. I point out, too, some of the mistakes that are possible if the task of writing a thesis is not approached in the right way. My examples are based on the students, like Brian, that I have worked with for several years each.

Completion of a thesis, especially a PhD thesis, involves mastery of a range of technical accomplishments, from learning an appropriate writing style to managing references, and from developing techniques for writing quickly to being effective at self-criticism and at criticizing the work of others. There is also the basic issue of learning what a finished thesis should look like. This book is structured as a discussion of the components of a thesis, and of the sequence of tasks you need to complete to get the thesis finished. The emphasis is on what you need to learn in order to do these tasks well, rather than on technicalities; other resources, including excellent books and websites, can provide help with different aspects of producing a thesis.

Using This Book

Chapters 1, 2, 3 and 4 concern how to get started, and what decisions to make before you even begin. Chapters 5, 6, 7, 8, 9, 10 and 11 show you how to tackle the various parts of a thesis and bring it to the point of submission. As a developing researcher, as well as writing a thesis you are probably presenting your research in journals and conferences, perhaps in collaboration with your colleagues or supervisor, a topic considered in Chap. 12; in this chapter I also consider some of the other challenges of being a PhD student.

I have used versions of this book as a source for graduate seminars and workshops on thesis writing. Those who are well into their writing seem to get immediate benefit from it. However, if you are at an early stage, I suggest you first read Chaps. 1 and 2 and—although this may seem surprising—Chap. 12. Some of it may not take on an edge of reality until you are well into your writing. As you will see, a key piece of advice (I would love to make it a command!) is that you start writing as early as possible, right at the beginning of your candidature. So you should also read Chap. 3, and get a sense of how best to make use of a word processor for authoring of a thesis, and of what the technicalities of thesis writing are. Make sure that you check the chapter summaries, which in some cases include discussion of useful kinds of online resources.

A book of this kind must navigate the variations in terminology and spelling between institutions and countries. I've had to make choices that might seem contentious, but to me the important thing is to be consistent. For example, I've chosen *program* instead of *programme*; *degree* instead of *program* (in another sense of the word); *graduate* rather than *postgraduate*; *thesis* rather than *dissertation*; British/Australian rather than American spelling (with the exception of the suffix '–ize'); *supervisor* rather than *advisor*; and *PhD* rather than *doctorate*.

Contents

Chapter 1
What Is a Thesis?

Simply defined, a thesis is an extended argument. To pass, a thesis must demonstrate logical, structured, and defensible reasoning based on credible and verifiable evidence presented in such a way that it makes an original contribution to knowledge, as judged by experts in the field. Among the many types of scholarly productions, theses are an oddity: each one is different, and there are no standard or generic constructions. Most of those who supervise theses have written just one, and, despite the effort they take to produce, the only people who carefully read a given thesis are the project supervisors, the examiners, and an otherwise rather select audience of specialized academics.

From the start, it is good to have a solid idea of what a thesis *is*, and perhaps the best place to start a discussion of theses is with their purpose. What do examiners look for when they judge your work?

Criteria for Examination

When universities send out a thesis for examination, they include their suggested guidelines for the examiners. I recommend that you get a copy of these guidelines from your own university (they are almost certainly available online) and look them over carefully. Make an effort, too, to understand the process of submission and examination.

At my university, the University of Melbourne < unimelb.edu.au >, the guidelines begin by listing key attributes of a successful thesis (quoted from the university's School of Graduate Research website, as of November 2010):

Attributes of a Successful Thesis

- The thesis demonstrates authority in the candidate's field and shows evidence of command of knowledge in relevant fields.

D. Evans et al., *How to Write a Better Thesis*, DOI 10.1007/978-3-319-04286-2_1,
© Springer International Publishing Switzerland 2014

- It shows that the candidate has a thorough grasp of the appropriate methodological techniques and an awareness of their limitations.
- It makes a distinct contribution to knowledge.
- Its contribution to knowledge rests on originality of approach and/or interpretation of the findings and, in some cases, the discovery of new facts.
- It demonstrates an ability to communicate research findings effectively in the professional arena and in an international context.
- It is a careful, rigorous and sustained piece of work demonstrating that a research 'apprenticeship' is complete and the holder is admitted to the community of scholars in the discipline.

At first glance these guidelines may appear to refer to the thesis, but they are really about the candidate. The first point makes this explicit: 'The thesis demonstrates authority in the candidate's field'. And consider the last point. The examiner has to consider whether the thesis 'is a careful, rigorous and sustained piece of work'—but see how it goes on—'demonstrating that a research "apprenticeship" is complete and the holder is admitted to the community of scholars in the discipline'.

At the start of introductory seminars in thesis writing, I ask students to explain the purpose of a thesis. Often they say something like, 'To tell people in my area about my research'. No doubt your research is of interest, but your primary purpose in writing a thesis is to pass an examination. These examiners are not reading your work out of mere interest: from the above criteria, we see that examiners read your thesis to assess whether or not you have demonstrated your fitness to be admitted to a community of scholars. Because a written thesis is an examination paper, not simply a report of research findings, you need to understand what examiners are looking for when they read your work. In the case of doctoral theses, examiners are encouraged to consider eight questions (quoted from the same website):

Guidelines for Examiners

- Does the candidate show sufficient familiarity with, and understanding and critical appraisal of, the relevant literature?
- Does the thesis provide a sufficiently comprehensive investigation of the topic?
- Are the methods and techniques adopted appropriate to the subject matter and are they properly justified and applied?
- Are the results suitably set out and accompanied by adequate exposition and interpretation?
- Are conclusions and implications appropriately developed and clearly linked to the nature and content of the research framework and findings?
- Have the research questions in fact been tested?
- Is the literary quality and general presentation of the thesis of a suitably high standard?
- Does the thesis as a whole constitute a substantive original contribution to knowledge in the subject area with which it deals?

These questions really are about the thesis rather than the candidate. They roughly parallel the structure of a solid thesis, and each builds on the previous one.

The first two questions are about familiarity with the previous work in your field and the demonstration of a critical approach to it. Note that, from the start, having and demonstrating a critical attitude towards your subject sets the tone of your interactions with the examiners.

The third question is about choosing appropriate research methods and justifying your choices as appropriate to the topic. Be aware that it is you, at this point, who must set the scope of that topic that will determine the appropriateness of a methodology. Further, the third question alerts examiners to show concern for the manner in which the methods are applied.

The fourth question focuses on displaying the results, explaining them and showing that you know what they mean. Here, then, it is not simply a question of showing those in the discipline area what you have found but also that you know how to present the results.

The fifth and sixth questions remind examiners to check the alignment, and connections, between an initial aim and the final conclusions. The logic flow in the thesis must be right. Notice, too, the emphasis on linking your interpretations back to what you said you would do earlier in the thesis.

The seventh question invites the reader to step away from the empirical side of the study to consider how well you can write. In a sense, the question signals to both you and the examiners just how important it is to be able to be able to communicate well within the international research community.

Finally, the eighth question asks examiners to consider the quality of the work as a whole. For doctoral students, producing work that is a 'substantive original contribution to knowledge' is a primary goal that can be reached through writing satisfactory responses to the series of previous questions.

There are other questions an examiner might also address. In particular, an examiner would look for evidence of insightful or critical thinking, and of objective appraisal of outcomes of the study. That is, they want to be persuaded that the student can think clearly and can construct a reasoned argument.

Types of Thesis

This book focuses on PhD study, but there are several other forms of research work that are understood to be theses. In the Australian context, the word 'thesis' is used to refer to the document that a student creates to earn a degree at the Honours, Masters, or doctoral level. (In other countries, such as the United States or Canada, the word 'thesis' is commonly used to signify work at the Honours or Masters degree level and 'dissertation' is generally used to refer to doctoral work.) What is the difference between the different understandings of a thesis?

At Honours level, a thesis—strictly, a 'minor thesis'—is a work of original research of approximately 10,000 words in length. For many students undertaking a minor thesis, it is the first time that they have conducted original research. From my experience, one of the main struggles occurs in making the transition from 'research consumption' to 'research production'. Minor theses are closely supervised and, very often, stem from research that is of direct interest to the supervisor. An Honours thesis is typically produced within a year alongside the demands of coursework. For the most part, they are assessed within the students' department; note, therefore, that the readership is well-known and thus the writing can be tailored to fit the audience.

At the Masters degree level, there are two types of theses. One is a minor thesis, with length limits ranging from 10,000 to 25,000 words. It is completed within one or two years alongside coursework, and usually requires one or two semesters of full-time effort. Much like those at the Honours level, minor theses are assessed within the department by a set of internal criteria.

The second type is a 'Masters by research' thesis of 30,000 to 40,000 words. It is much more substantial than those that are written by coursework students as it is the result of full-time research over one to two years. This thesis is examined by experts in the field outside the department. In some departments, students first join the field by writing a Masters thesis; if it is considered to be of high quality and can be extended, it can be converted into a doctoral thesis.

A 'Doctor of Philosophy' is earned by the successful completion of a PhD thesis. For PhD students, the word limit of a thesis is 1,00,000 words; most students write approximately 80,000 words. In Australia, a PhD thesis is typically produced in 3 years of full-time study. It is examined by two experts who have themselves supervised doctoral work; and they are likely to be located at an international research institution.

There are other types of doctorate, too, including those in education, by exhibition (in fine arts), or by publication, but these are beyond the scope of this book. All of these should be described in the policies on your university's website.

Look at Other Theses

It's now time to look at some other theses. Most supervisors have a few on their shelves that they may be willing to lend you. Reading these works will be a good start, but don't stop there. Probably they follow a pattern set by your supervisor's own ideas of a good thesis, and almost certainly they will be typical of what your own department thinks is acceptable. So go out and look at theses from across a range of disciplines, and even theses from other countries. As presentation and style change relatively rapidly, look at theses that are no more than 3 years old. If applicable, examine a mix of kinds of studies, both qualitative and quantitative (see Chap. 8). Try and find work that is outside your field, but makes use of a similar methodology. After you have skimmed several, select some that are coherent, and some that are not so clear, and go through a few of them with your supervisor.

Read the theses as if you were an examiner. With the guidelines for examiners in front of you, begin the assessment of each of them by first looking at the overall layout. See if the table of contents gives you a clear idea of the structure of the work as a whole. Then browse the introduction and conclusions, and look through the reference section. Next, read the introduction carefully and compare it to the conclusions to see if the work is linked in a coherent manner (see the fifth question in the guidelines for examiners on page 3). It might surprise you to find that some theses fail to make this link. Look especially for specific formatting and conventions: How are particular words spelled? What is the best way to display data? What is the typical length of a chapter? You may be impressed with the virtues of some theses, such as professional layouts, innovative displays of complex material in graphs or tables, or a strong integration of online materials. Stay alert for the points that impress you, and make a note to adopt them for your own work.

Examiners' Reports

Students are sometimes advised to track down examiners' reports on submitted theses. For the most part, the examination process is confidential, but make an effort to ask a completed student for a report or see if a supervisor is willing to share an examination that is anonymous. As you read examiners' reports, or the associated studies on them, get in the frame of mind of these expert assessors. What do they look for, and what do they ignore? Do they directly answer the suggested questions put forward by the university? These reports will be highly variable in detail and approach; What can you learn from these differences? Additionally, seek out academic studies that concern thesis examination (search for the keywords: thesis quality, doctoral assessment, research training, PhD examination) with a view to developing a better understanding of the assessment process. Feedback from examiners is summarized in the Appendix, which is a digest of observations from examiners' reports.

I have examined numerous theses of each type: minor, Masters, and doctoral. In each case, my purpose is to assess the work with reference to the criteria at hand. My considerations vary. At times I focus my comments on the big picture; at other times I hone in on details. My motivations for examination are not necessarily to hand out criticism, or even praise, but to sharpen a study. Academics examine theses partially out of service to the profession and partly as a favour to those who ask, but mostly to learn something new before the work is presented at conferences or published in journals. In short, as an examiner, I am looking to learn and, in this way, I'm just like the candidate.

Consider the five potential outcomes of PhD examination at my university (edited slightly for clarity) that an examiner can choose from:

- Be awarded the degree of Doctor of Philosophy without further examination or amendment.
- Be awarded the degree of Doctor of Philosophy without further examination, subject to inserting in the thesis the minor corrections or additions as specified

to the satisfaction of the Chair of Examiners, without further reference to the examiner.

- Be awarded the degree of Doctor of Philosophy subject to revising part or parts of the thesis to the examiners' satisfaction.
- Not yet be awarded the degree, but be permitted to resubmit the thesis in a revised form for re-examination. Areas requiring major amendment are identified in an examiners' report.
- Not be awarded the degree of Doctor of Philosophy and not be permitted to submit for re-examination.

Think for a moment of the implications of each of these outcomes. Remember, first of all, that there are two expert examiners who are assessing the work. If a student is awarded a PhD because both examiners have marked that it fits the first criteria, no more amendments are required. Nothing, not even occasional typos, requires change: the only thing left to do is to make a bound, final copy to be archived at the university, and perhaps submit an electronic copy to be placed online.

Many students (including myself) earn the second mark. That is, they have been awarded the PhD, and no further examination is required, but there is a need to make some corrections, write out a report to the Chair detailing the required changes, and reprint the thesis for submission. By awarding a PhD based on the third outcome, an examiner indicates that the student must revise entire sections. Substantial work is required, and the revised and reprinted thesis must be sent back to the examiner for checking. The use of the fourth mark by an examiner indicates that the thesis requires such major revisions that a PhD cannot yet be awarded, but the work can be re-submitted. Finally, on occasion, examiners use the fifth outcome to deny both an award of a PhD and a chance to submit a revised thesis.

Examination processes for minor theses are highly variable, with students being awarded pass/fail in some cases or a mark in others. Some processes allow for resubmission; some do not; examination may be within the department. In many institutions Masters theses are handled in the same way as PhDs, but in some places different processes are used. Make sure that you are familiar with the mechanisms that apply to your degree.

Summary of Chapter 1: What Is a Thesis?

On theses:

- There is no 'standard' definition of a thesis but it is generally understood to be the result of structured, original research that is produced for assessment.
- The expectations for a thesis vary from university to university, field to field, and supervisor to supervisor.
- There are several types of theses in the range of research higher degrees. Some theses are produced alongside the demands of coursework, and others fulfil the total requirements of the degree. The types of thesis vary in length, complexity, comprehensiveness, and even purpose.

On examination:

- You need to understand the criteria for examination of theses, and be sure to craft your own work so that it meets these criteria. Be familiar, from the start, with the attributes that are expected of student candidates.
- It can be rewarding to read and analyze theses both from your own field and across other disciplines. Note weaknesses that you wish to avoid, and strengths that might be adapted for your own work.
- A summary of examiners' responses is included as an appendix to this book.

Online resources:

- There are numerous online indexes of theses and dissertations. For example, many Australian and New Zealand theses are available at the National Library of Australia's website, or through individual university library collections.
- Your university library should provide access (in paper or online) to all of the university's PhD theses.
- Policies for examination, and descriptions of thesis types, should be on your university's website. You should also browse your university's policies and procedures that relate to research candidature.

Chapter 2
Thesis Structure

Karen was undertaking a PhD in engineering to investigate whether a new type of plastic was safe to use as cookware. When she started her lab work, she decided to begin writing her thesis, but despite her determination she was having trouble. I knew Karen well, and she was a very good student who had been interested in new plastics ever since her undergraduate studies several years ago.

Karen decided that the first thing to do was to write a review of the literature. I told her to send me an outline of how she expected to tackle it soon—but after two weeks nothing had yet appeared. I asked her what the problem was. 'No problem', she replied, 'I just have a lot more papers to read. When I've read and summarized them, then I can start writing'. I reminded her that she'd told me a similar story a few weeks earlier: after reading a few more articles, she would indeed start writing. While reading those, however, she'd turned up several more. And then there was the material that she had listed to read in the future. Karen then showed me several summaries, and each was separate. At that point, I concluded that she was never going to start. Seemingly, Karen had told herself that finding 'a few more papers' was the reason for her continued delay, but from experience, I realized that her problem lay deeper.

Why We Have Trouble with New Tasks

When we start a new project, figuring out how to proceed is easy if the project is similar to things we've done in the past. Building a bookcase, say, is not a big challenge for someone who has already made a kitchen cupboard. But an entirely new task is another matter: every aspect is unfamiliar, and it is not obvious how to begin or what the obstacles will be. We may not even know how to think about the problem. Imagine the state of mind of someone whose practical experience is limited to building kitchen cupboards, but who is asked to build a three-bedroom house. There will be many questions: Where to buy the materials? What materials? What tools are needed? Will the walls be strong enough? How to arrange for plumbers and plasterers? What is the first step? The task of starting to write a thesis may be equally as challenging.

D. Evans et al., *How to Write a Better Thesis*, DOI 10.1007/978-3-319-04286-2_2,
© Springer International Publishing Switzerland 2014

Research is unpredictable. In nearly every project I've been connected with, the conclusions contained some unexpected elements. In most projects the aim of the work changed as it progressed, sometimes several times. I've often—startlingly often!—had students say that their 'experiments had failed', but, when we had absorbed the implications of the supposed failure, new hypotheses emerged that resulted in breakthroughs in their research. On several occasions truly surprising conclusions were staring the student (and me) in the face, yet we failed to see them for weeks, or longer, because we were so hooked on what we *expected* to find. That is, continuing the analogy above, we may not even be sure of what kind of building we are trying to construct.

Moreover, the process of research is often not entirely rational. In the classical application of the 'scientific method', the researcher is supposed to develop a hypothesis, then design a crucial experiment to test it. If the hypothesis withstands this test a generalization is then argued for, and an advance in understanding has been made. But where did the hypothesis come from in the first place? I have a colleague whose favourite question is 'Why is this so?', and I've seen this innocent question spawn brilliant research projects on quite a few occasions. Research is a mixture of inspiration (hypothesis generation, musing over the odd and surprising, finding lines of attack on difficult problems) and rational thinking (design and execution of crucial experiments, analysis of results in terms of existing theory). Most of the books on research methods and design of experiments—there are hundreds of them—are concerned with the rational part, and fail to deal with the creative part, yet without the creative part no real research would be done, no new insights would be gained, and no new theories would be formulated.

A major part of producing a thesis is, of course, creating an account of the outcome of this rational–creative research process, and writing it is also a rational–creative process. However, the emphasis in the final product is far more on the rational side than the creative side—we have to convince the examiners with our arguments. Yet all of us know that we do write creatively, at least in the fine detail of it. We talk of our pens (or fingers on the keyboard) running ahead of our brains, as if our brains were the rational part of us and our fingers were the creative part. We tend to separate one from the other. Of course this is nonsense, and we know it, yet the experience is there.

Wrestling with this problem has led me to the view that all writing, like all research, involves the tension between the creative and the rational parts of our brains. It is this tension—as well as our lack of experience in the specific task of writing theses—that makes it so hard for us to start writing, and sometimes gives us 'writer's block'. To get started, we must resolve the tension.

Structuring Your Thesis

A colleague was concerned about the draft thesis that had been submitted to him by Henry, one of his students, and asked me to look at it. It was certainly difficult to know what was going on. Henry had written the draft straight from a logbook, experiment after experiment, in chronological order:

Experiment No. 37: as Experiment 36 failed to show the chemical reaction I expected, I next tried the effect of doubling the concentration of the active reagent ...

... and so on. In other words, Henry had presented a condensed diary, which certainly detailed the work he had undertaken but lacked the essential elements of a thesis: motivations for decisions made, interpretation and explanation, linking of data to conclusions, and argument supporting propositions and hypotheses. Your task as a writer is to document your processes, but equally to make these processes and the outcome of your work comprehensible to readers—not to explain how you spent your time, or to describe the hypotheses that ultimately didn't make sense. You need to structure your thesis in such a way that you take the reader from the aim to the conclusions, via the evidence and arguments, in the clearest possible way.

As noted, there is no such thing as a standard thesis, but a careful reading of the guidelines for examination does suggest that there is a standard thesis *structure*. In essence, a thesis must first motivate the study, present background material and conduct a study. Results must be well argued and displayed, and the thesis has to end with a sound conclusion. My experience is that this standard structure works well for theses in the physical, biomedical, mathematical, and social sciences. The nature of research in the humanities is different from that in the sciences, and different forms of reportage may be appropriate for theses in different areas.

The 'Standard' Thesis Structure

The standard thesis structure has four parts: an *introduction*, the *background*, the *core* (for want of a better word), and a *synthesis*. Note how, as illustrated in the following figure, the sections are connected to each other. A conclusion responds directly to an aim, for example, and the background must directly foreshadow the core.

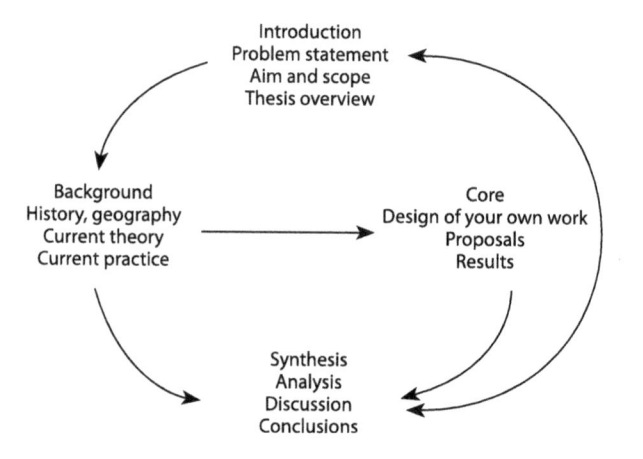

Some of these parts might contain more than one chapter, and the core might be more than half the thesis. Each of these parts has a distinct role.

The *introduction* explains what the thesis is about: the problem that the thesis is concerned with, the aims and scope, and the thesis structure. In some disciplines it includes an overview of the findings. An introduction is typically written for a wider readership than the bulk of the thesis, and may use illustrative examples to help underpin the reader's understanding of what you are trying to achieve. Such examples help to create a narrative that a reader can use as context for your work. However, an introduction isn't an essay—the only purpose it has is to introduce the research. You should outline the problem you have investigated, explain the aim of the research and any limits on the scope of the work, and then provide an overview of what lies ahead. Five to ten pages is ample.

The *background* is the knowledge required before a reader can understand your research: relevant history, context, current knowledge, theory and practice, and other researchers' views. In the background, your purpose is to position your study in the context of what has gone before, what is currently taking place, and how research in the area is conducted. It might contain a historical review. If the research is location-specific (an investigation of diet in low-income suburbs, for example, or an examination of how a dialect is changing) you will need to describe the study area and its characteristics; if the research is technology-specific (such as a study of food packaging or the yield of a harvesting machine) you will need to describe the specifics of this technology and how it affects the questions you can ask. The background usually contains a chapter reviewing current theory or practice, and may include the results of preliminary experiments or surveys carried out to help you feel your way into the problem. Experiments may also be used to establish benchmarks based on other work against which your work is to be measured, and these too form part of the background.

The *core* concerns your own work: your propositions or hypotheses, innovations, experimental designs, surveys and reviews, results, analysis, and so on. (This is sometimes called the contribution, though in a strong thesis the background too forms part of the contribution, as other researchers may value your interpretation and analysis of past work as much as they value the 'new' work presented in the core.) The core can easily form the bulk of the thesis and consist of several chapters.

The *synthesis* draws together your contribution to the topic. It will usually contain a discussion in which you critically examine your own results in the light of the previous state of the subject as outlined in the background, and make judgments as to what has been learnt in your work; the discussion may be a separate chapter, or may be integrated with the detailed work in the core. Finally, it is where you summarise the discussion and evaluation to produce conclusions. These should respond directly to the aim of the work as stated in the introduction.

The structure of the core varies greatly from discipline. In one thesis, the first of the chapters in the core might be a description of a survey tool and an explanation of how it is linked to an investigation of why obese people make poor dietary choices; the next might be a presentation and statistical analysis of the results; and the next two chapters a presentation of a detailed study of a small number of individuals, looking at the impact of methods of changing their behaviour. In another thesis, the first core chapter might sketch why it is plausible that a particular food has an effect on the immune system; the next might propose specific chemistry that would cause this effect; the next might describe an experimental design to test for this chemistry; and so on.

A common factor is that the core is a narrative leading from a proposition to an outcome, linked by evidence and argument. In a more complex thesis, there may be a series of linked propositions, each independently supported by evidence and argument. I return to this issue in Chap. 7.

Below is a typical application of this structure, for a thesis examining the role of labels in diet choices. The thesis has three background chapters, which examine two aspects of labelling—legislative requirements and marketing—and social issues around food choices. These insights are used to develop a research survey for identifying the level of understanding of and belief in labels, which in turn is used to propose and test the impact of alternative labelling mechanisms.

The Influence of Food Labelling on Young Adult Diet
Chapter 1 Introduction
Chapter 2 Food Labelling Legislation
Chapter 3 Food Marketing Strategies
Chapter 4 Factors in Young Adult Choices
Chapter 5 Research Method
Chapter 6 Comprehensibility of Food Labels
Chapter 7 Alternative Label Designs
Chapter 8 Identification of Effective Labelling Factors
Chapter 9 Discussion
Chapter 10 Conclusions

These four parts (introduction, background, core, and synthesis) are examined in detail in Chaps. 5–10. My aim in this chapter is to convince you that you should ensure that each of them is progressively developed as your writing proceeds. A strong thesis is the product of considered work, where there has been opportunity to debate, revise, and evaluate each chapter at leisure; and is particularly strong if the components are tightly integrated. This integration is most easily achieved if they are written concurrently.

How many chapters should a PhD thesis have? If there are four main parts, each containing one to three chapters, we should not expect more than eight or ten chapters altogether. Many theses are accomplished in five to seven chapters. If you have more, you should suspect that some are really only sections, and need to be consolidated. In some disciplines, theses are assembled by editing papers that the student has published during the candidature to produce a coherent whole. With careless editing, such an approach can easily lead to a series of brief or fragmentary chapters that don't form a consistent and sustained argument; that is, the collection cannot in itself be considered a thesis. In far too many cases some of these chapters are preliminary work, or work that it is off-topic, that shouldn't be included at all. My experience is that producing a thesis in this way is usually much harder than the student expects—often the student feels that, since the papers were published, they are 'finished' and all that is required is to gather them together—whereas even an experienced writer needs at least 5 or 6 months to turn a set of papers into an acceptable thesis.

My university once asked me to report on a request for financial assistance to publish a thesis as a book. It had around thirty chapters! The simple and coherent structure discussed above was totally obscured by the proliferation of chapters with seemingly arbitrary titles. The effect was total fragmentation of the reasoning and impact, and I was surprised that the examiners had passed it.

Narrative

One way to think of the role of structure, and signposting, is as a kind of guide that walks readers along a road from what they did know (past knowledge) to what they should know (a knowledge frontier). When you write a thesis, it can be helpful to reflect on what you knew—and how you thought—when you began your work. This earlier 'you' is the person you are writing for. The story, or narrative, that takes the reader along the road should be as straightforward as you can make it. That is, you may think to yourself: I have had to fumble, and explore, and make mistakes to get here, but I am now writing the guidebook that helps the next person to painlessly come to the same point of view and the same knowledge.

A key element to good writing is to clearly understand what the writing is meant to achieve. In my view, the twin concepts of narrative and audience—what you are trying to say, and who you are saying it to—are the most important lessons a writer can learn.

Look for the structure behind the material you are describing, and don't confuse narrative with structure. The narrative concerns how you want the reader's thoughts to develop as they read the thesis. The structure is how the material is organized to create a narrative. Different structures may be appropriate in different areas, particularly between the humanities and (in the broadest sense) sciences. In an empirical study, the structure might be: the problem and its significance; relationship to previous work; derivation of hypotheses; design of experiments; results; analysis and interpretations; conclusions (with, perhaps, two series of experiments, the second resting on the outcomes of the first). In contrast, in a literary study the structure might be: the purpose of the study and its contribution to knowledge; evaluation of previous studies; procedures, limitations, and assumptions; sources and documentation; analysis of facts and evaluation of evidence; conclusions. These structures are not identical, but there are strong similarities.

There are other differences between theses. In some disciplines, it is the norm for a thesis to be a consolidation of several papers; in others, the thesis is usually a single large piece of work. Some emphasize quantitative work, with, in the extreme, a thesis where the contribution is mathematical theories or lab experiments that lead to precisely quantifiable outcomes; others emphasize qualitative work, with, for example, discussion and argument based on documentary sources and other researchers' interpretations of records of events.

Something that all theses have in common is the need for analysis and reflective consideration of the issues. Too often, researchers run the risk of merely describing their complex settings and ignore the need to demonstrate critical thinking.

Non-standard Thesis Structures

Some theses do not fit into a standard structure. Across a wide range of disciplines there is a trend towards a blending, for example, of quantitative and qualitative approaches. Such work might include, for example, an in-depth examination of the context and history of a situation before arriving at a 'statement of the problem'. A quantitative survey might inform the development of interview questions, and these in turn might lead to analyses of the results that may suggest yet another series of questions. A series of chemical experiments may be inspired by a revisiting of a historical dispute, and be built on an analysis of arguments for competing methodologies. A conceptual framework may be an outcome and not a starting point.

If you are writing a thesis that relies on a non-standard structure—or are writing a thesis where the approach and problem might, in traditional terms, be 'interdisciplinary'—don't make the mistake of trying to reinvent the form of the thesis from scratch. Take the time to find other theses that have pursued similar problems in a similar way; read these theses, and others, to help yourself decide how your work should be organized and presented. Make sure you are familiar with the methods of both qualitative and quantitative research; there are many excellent books on these topics, some written for specific disciplines but readable by a broad audience, such as the books on statistical research methods for psychology. And it is essential that you establish a clear line of argument throughout your work.

As discussed in Chap. 1, be sure that you know the criteria for examination. Just because you are doing something 'different', you are not excused from creating a strong academic argument that is underpinned by sound evidence, credible analysis, and clear writing. How to use these elements in creation of a strong thesis is the subject of the next few chapters.

Summary of Chapter 2: Thesis Structure

Your thesis should be organized as follows.

1. An Introductory Chapter

   - Tell the reader the problem you are tackling in this project.
   - State clearly how you aim to deal with this problem.
   - Limit the scope of your study.
   - Sketch out how the thesis is structured to achieve your aim.

2. Background Chapters

  – Include in these chapters all the material required to lead up to your own work.
  – Ensure that there is a flow of narrative that explains why each topic is being discussed.

3. A 'Core' Account of Your Own Work

  – Begin with a formal statement of your hypotheses or research questions.
  – Follow this with an account of the methods you chose to test your hypotheses or answer your questions, and why you chose them.
  – Report the results of applying these methods.

4. Synthesis

  – You are now ready to pull the whole thesis together.
  – Discuss the implications of your results.
  – Draw strong conclusions backed up by your discussion.
  – Check that they respond to the aim stated in your introduction.

Things to consider:

- Are you are blocked in your writing, or procrastinating? Do you understand why? If not, discuss it with someone.
- Think about how your thesis will work as a narrative.
- Decisions about organization should have a rational foundation. Satisfy yourself that you have good reasons for your chosen thesis structure.

Chapter 3
Mechanics of Writing

Given a structure, the next challenge is to actually begin writing. Strategies for writing are the subject of the next chapter. In this chapter, I take a brief detour and consider some of the tools of writing and communication.

Writing at a Computer

Most people do all their own typing and word processing—a dramatic change from just a decade or two ago, and nowadays the old approaches to thesis writing (which included use of professional typists and a great deal of writing by hand) are almost forgotten. All students are familiar with the challenges of using a word processor to produce a well-written short work such as an essay. They are also familiar with the modes of writing that computers encourage: frequent revision, writing of sections in any order (or of several sections at the same time), ease of change of style and layout, and so on. The use of word processing has become universal, profoundly affecting the way research can be carried out and reported.

However, the task of writing an extended document such as a thesis is a very different process to that of writing a shorter work. Many students know the elementary features of word processors that are sufficient for a 3,000-word essay, but not the more advanced features that help authors to maintain consistency of style and presentation over 50,000 words or more, a scale on which manual checking can become painfully laborious and where it is essential to have automatic maintenance of elements such as section and figure numbers. Likewise, good presentation requires software that automatically maintains bibliographies; mechanisms that create indexes and tables of contents; tools for professional-looking illustrations; and strategies for keeping versions and back-ups.

On the other hand, some students tend to over-rely on some of the functions of word processors, such as grammar and spelling checkers, which may be designed for general writing rather than the demands of communication within a specific academic community. Learning to make effective, appropriate use of a word processor is a key step towards timely completion of your thesis, and it is a mistake to assume

D. Evans et al., *How to Write a Better Thesis*, DOI 10.1007/978-3-319-04286-2_3,
© Springer International Publishing Switzerland 2014

that even long familiarity with a particular word processor means that you are using it well as a research tool (or that it is the right choice for your new activity).

Today, the most widely used general-purpose word processor is Microsoft Word, or the OpenOffice equivalent; in the mathematical and physical sciences many researchers use the more technically oriented markup-based LaTeX. I do not explore the specifics of these word processors, but encourage you to use resources such as advanced guides and manuals to ensure that you are using them well—even an occasional revisit to an online tutorial can be surprisingly rewarding.

A typical word processor can be viewed as a suite of separate tools, perhaps bundled together under a common user interface. These tools might include an editor, for entering and modifying text; spelling and grammar checkers; a bibliography database; a line-art environment; and a system for laying out the text in a form suitable for printing or for viewing online. This last point is particularly important: a word processor allows the *style* of a document to be separated from its *content*, and the two issues are important at different stages of the thesis creation process.

Presentation

The fine details of the style of your thesis may not become settled until it is almost finished, and I strongly encourage that, in the early stages at least, your focus is on creation of content rather than on how it appears on the page. Nonetheless, right from the start you need to write within the constraints of a style—for example, so that all headings of the same kind, such as chapter titles, are displayed in the same way.

Which format do you adopt? In principle, you have great latitude in your choices, so long as you are consistent in what you do. In practice, however, you should adopt a thesis style that is already in use in your department, and it may well be that your supervisor has specific requirements. You should also be aware of field-specific guidelines, such as those published by the American Psychological Association. That said, some standard styles are less than eye-pleasing, and you may want to refine the appearance. The key thing is that you use a style from the start, so that you can easily change the appearance of the whole thesis if you want to do so.

Pay particular attention to the way you cite references, both in-text and in the bibliography or reference section of your thesis. Your professional handling of references is one way that examiners assess your readiness to enter the community of scholars. If you are sloppy, or maintain incomplete lists, or perhaps fail to cite a work, it signals that are you not respecting colleagues. Quite apart from the inherent importance of this, you will annoy an examiner if you cite material and fail to list it. One way an examiner checks to see whether you know what you are talking about is to check the references as you cite them. Conversely, you shouldn't put references in your list of references *unless* you have cited them. So all of these have to be checked, one by one. Read your own text the way that the examiner would, checking the list every time you come to a citation. My advice is to be systematic when you are collecting the reference material in the first place, and remain aware of the importance of correct citations throughout your professional career.

Using a style means that, once you have established a pattern, you can easily stick to it and the reader will get the same message every time. For example, main section headings, wherever they appear in your document, will always be in the same typeface (font) and of the same size. They should always be preceded by the same space separation from preceding text, and always be followed by the same space separation. If the style you choose is clearly different from that of other headings, the reader will quickly understand 'We are starting a new main section' or 'This is a sub-section within the section'. This is especially true if you use a style that is familiar to most readers in the field.

A thesis consists of several different parts that need to be tied together with a set of conventions. Without a standard format across the entire document, the work will appear random and unprofessional. For example, you should put all chapter headings on a new page, using the same style; that is, the same font and paragraphing. You should give all major section headings a style that is different from that of the chapter headings. Captions to figures should all have the same style, but be different again from section headings and different from the main text. All new paragraphs should begin with the same indentation (except for the first paragraph after a heading, which may have no indent at all), and so on. All this will help your readers to navigate their way through your thesis. This styling is provided with templates, which govern the appearance and numbering of every element of a document.

Before you start writing your report or thesis, you should think about its format and devise styles and formatting rules that are appropriate for your field of study. Begin as you mean to continue. Introduce rules as necessary, and be aware that too much complexity can work against you. For example, avoid deep structures—is it really necessary to have paragraphs with numbers like 3.1.2.1a(iii)? Once you have a style, any element of the document can be put in that format, and you are on your way to producing a professional-looking thesis. While you may have had little previous exposure to creation or use of styles, in my view templates are the single most important feature of a word processor, and you *must* learn how to use them properly.

After creating a style, you can generate a thesis structure, with a few empty chapters and perhaps some subheadings and so on. You can then use the style to generate a table of contents, and begin to get a sense of how the final thesis will appear. As you proceed, you will use the table of contents, or other outline tools, to get a sense of the current structure of the thesis and where it may need revision—extra chapters, moving of material from one section to the next, changing how headings relate to each other in the hierarchy, and so on.

Writing Tools

Most word-processing programs include a facility for checking spelling. It checks every word you have typed against a dictionary built into the program. Do not ignore it![1] Few people are infallible spellers or proofreaders and it is only rational to

---

[1] It's been argued that spell-checkers make authors lazy, and that writers at any level with access to a spell-checker make more mistakes than those without. I suspect that this is a case where average

have oddities questioned. However, although the spell-check is good at picking up typographical errors, it can't make decisions for you. Typical problems are proper names (people's names or place names), and words for which there are alternative spellings. In the case of proper names, the temptation is to tell the program to ignore its questionings, and go instead to the next area of doubt. This is a mistake: you should check any proper name the first time the spell-check comes to it and, when you are satisfied that you have got it right, add it to the dictionary. The second problem is words for which alternative spellings are permissible (*-or* or *-our* and *-ize* or *-ise* are the most common). The most important constraint here is that you be consistent. Before you start, determine your preferred spellings for these words, and keep to them.

On a related note, don't rely on the spell-check to proofread for you; although it will pick up misspelt words, it won't distinguish between, for example, *there* and *their*, or *affect* and *effect*. Similarly, it won't tell you if you have left a word out.

Grammar-checkers look at every sentence, and make checks such as: Does it contain a verb?; Is it missing connecting words?; Does the subject agree with the verb (plural subjects must not have singular verbs)?; Is the verb in the passive voice (permissible, but should be used sparingly)?; Are stock phrases being used (examples: 'over and above'; 'in order to', 'part and parcel')?; and so on. You may think that your English is better than that of the grammar checker—and some of the time you will be right—but my experience is that they are often useful, and it is essential to use a grammar-checker at least once before finalizing your thesis.

References

The *name and year* (or *Harvard*) system is the most popular reference style for theses. It works well for readers, because it names references in an understandable way, and also works grammatically. We can write for example 'Rami and Tuntara (2002) found little evidence of ...' or '... there was little evidence (Rami and Tuntara 2002)'. There are many alternatives, but the Harvard style is simple and effective.

Another form of referencing is the *numbered note*, or footnote, system, which is used in many books. Such notes are usually collected at the end of each page. When the page is printed out you will see the superscript number in the main text and the footnote text at the bottom of the page, separated from the main text by a dividing line. If you have more than one reference number on the page, the footnotes are all collected automatically on that page, or instead the notes can be collected at the end of a chapter or in a consolidated listing at the end of your thesis. Note that tools for maintenance of references provide mechanisms for changing from one style to

---

behaviour is meaningless—for some people it is an essential tool, while for others it is little more than a distraction. However, spell-checkers are at best a partial solution, and do not allow an author to avoid the task of a thorough proofread.

another, but, if you do make such changes, make sure the sentences with citations still parse correctly.

Wherever you put them, the notes have to be backed up by a consolidated alphabetical listing of all the references. A typical PhD thesis will end up with two hundred or more references (yes, you will read that many, and understand them). Even a minor thesis may have thirty to sixty references. Keeping track of these is a daunting task. For this reason alone it is worth learning how to use an effective bibliography tool. Word-processing programs can collate and maintain references, using bibliography software that builds up a catalogue of references. Each entry consists of elements such as, in the case of a journal article, the author (or authors), the title of the article, the journal name, year of publication, and publisher and place, together with an optional abstract and keywords. This reference database can be used independently of your thesis as a way of recording the papers you have read, where they can be accessed, and your views or comments on the content.

Whichever system of referencing you use, the word processor offers the advantage that it helps you to maintain the match between the references cited in the text and the references appearing in your consolidated bibliography. It helps to prevent you from inadvertently omitting references from your list that have been referred to in the text, and also helps to prevent you from retaining references in the list that are no longer referred to.

Make sure you capture full bibliographic details, and perhaps a permanent URL such as a DOI < doi.org >. I recommend keeping a softcopy (that is, an electronic version) of all papers that you find online. You will end up with hundreds, so be sure to organize them carefully. If you really feel that you need paper versions to help keep you organized, consider strategies such as just printing the first page.

Tables and Figures

Word processing software includes rich mechanisms for assembling tables, whether of numbers, images, survey responses, or some other data. It is up to you to ensure that your tables have an obvious logical structure that readers can easily understand, and it also up to you to ensure that you make good use of the software, not simply use the defaults.

Consider the breadth of uses of Microsoft Word: school children drawing pictures; managers dashing off memos; journalists typing up articles; clubs producing membership lists; and on it goes. It is hardly surprising that the default settings aren't particularly well suited to the specific, niche task of thesis writing, and yet in many theses the author has made no effort to improve the look of the work. Tables in particular often seem to be poor, with upper case headings as if the author is SHOUTING, bad vertical alignment of values in the columns, illogical and inconsistent organization, and heavy lines everywhere. Such tables are a sad contrast with the presentation in a typical professionally typeset journal, and yet the effort to turn one into the other may be only a few minutes.

I recommend that you find good models and imitate them. Remember too that tables are sometimes copied and used out of context—in slides, lecture notes, and so on—so they should be reasonably independent of the text. That is, take the effort to create captions and headings that make them at least somewhat comprehensible on their own.

Similar comments apply to figures, that is, graphs and artwork. (Illustrations are discussed further in Chap. 8.) If your word-processing package has a reasonably sophisticated graphics or line art system, you might consider using it to draw your figures. This method has the advantage that you can edit the figures at a later date in the light of rewriting or alteration of the text. However, you will have to accept the limitations of the inbuilt graphics system, which may be significant. Alternatively, you can draw all your figures using a separate specialized graphics package, and import them electronically into your text. You can't edit them while they are in your word processor, but you can delete them, go back to the original version in the graphics package, edit that, then re-import.

If you are using charts generated by a spreadsheet program or a statistical package, these too can be imported into your text. If you want to plot your data, enter it[2] into such a package rather than attempting to create a graph with a line art tool—such plots are amateurish at best—and take advantage of the facilities that these packages provide for displaying data in a variety of useful formats and perspectives. Don't assume that the standard layout provided by the software is going to be best for your data; you may want to use colours, different kinds of labelling, different kinds of ways of representing quantities, and so on.

Whichever method you use, ensure consistency of style within the figures, especially if there is written material actually on the figures themselves, such as labels on the axes of graphs. You should produce the captions of all your figures, together with any explanatory material and references to sources, in a consistent style, using your word-processing package. Note that it is customary to put the caption to a table above the table, but to put the caption to a figure below it.

There are numerous texts on how to present figures and diagrams; Edward Tufte's books are exemplary. There are also good online resources. For line art, there seems to be a widespread attitude that quick enough is good enough. Where once the diligent researcher had to rely on the skills of a graphic artist to produce suitable illustrations, now anyone, including small children, can access simple line art tools. Used carelessly, the result does indeed often look like the work of children, from cute[3] use of inappropriate clipart to lack of logical organization. It is like people believing they are producing works of art because they take family snapshots with expensive cameras.

Your tables and figures are intended to convey the key messages of your thesis, so give them the time and care that such messages require. I continue to be astonished, and not in a good way, by the fact that students who labour for months or

---

[2] Yes, data is an 'it'—not a 'them'—in this context, because I am not referring to a numerable quantity, or a set made up in a meaningful way of individual datums [sic], but to a form of information. Pedantry may be with 'them', but usage isn't.

[3] Here I use 'cute' in the sense that is closer to nauseating than to sweet. Please take the hint.

longer over the text of a thesis often seem satisfied with throwing together illustrations in a few minutes. In the process, they produce ugly or misleading figures that undermine their work and erase any impression of quality. Take the time to locate best-practice models to imitate, and work over your figures until they are as good as the rest of your work. As is true for text, producing a good illustration takes time, and includes drafting, reflection, revision, and iteration.

In some older theses, all the figures and tables were collected together at the end of each chapter, each on a separate page. This was a product of the typewriter age; you won't find them like that in a book. The word processor enables you to enter them in much the same way as in a book: as close as you can get them to the place where they are first mentioned in the written text.

Tracking Changes

Some word processors have facilities such as 'track changes' commands. These allow you to alter a document, see what you have done, and later decide whether to 'accept' or 'reject' your decisions. It can be used collaboratively for documents with multiple authors. Take note of it and learn to use it but not to excess: often, students worry too much about every word they have typed to the detriment of their creative processes. Nor is tracking changes an effective method of version management, that is, explicit storage of drafts and versions as the work progresses (this is discussed in the next section).

Some word processors have another useful feature—the ability to selectively hide text so that it is visible during content preparation but not in the final print or online format. This text can be used by authors for comments or reminders, or as a way of keeping text that seems unnecessary but might be useful in another context—remember, for example, that some text might be used in both a chapter in a thesis and a paper that is prepared for separate presentation. Again, this feature should not be overused; I've seen theses in which most of the text was 'hidden' and the student could no longer read or edit drafts effectively. But it is powerful when used well.

Document and Version Management

Before the advent of word-processing, thesis drafts were hand-written or typed. Photocopying was expensive and, as a result, drafts were precious. The nineteenth century author J Meade Faulkner left the only copy of one of his four novels on a train, and it was never recovered. In principle, such a loss seems inconceivable today, but the use of word processors has spawned a completely new set of ways for work to be lost, mostly to do with poor version or document management—but also because of a misplaced sense of trust in the reliability of computers. The great slave can be an obstinate enemy! If you follow a few simple rules you will avoid the most common problems.

- Frequently save the document you are working on. Most editors have a feature for automatically saving to disk every few minutes. This copies your file to whatever working storage you are using—local hard disc on your desktop, your area on a departmental file server, or a home server. In addition, I strongly recommend that you copy your work to some other backup storage every hour or two.
- Backup storage technology for home computers continues to change; I've used cassettes—yes, home audio cassettes were once a storage medium for personal computers—four different kinds of floppy disk, four different kinds of digital tape, CD-ROM, DVD, external hard drives, and at the moment, mostly, USB flash memory. These are all examples of non-volatile storage, that is, the data isn't lost when the power goes off. USB flash drives seem to be particularly robust, and at (in 2011) a few dollars per gigabyte, a student can easily buy several and use them in rotation. Leave a USB drive plugged into your computer, and drag your files onto it every now and again.[4] Make sure it always has the most recent or 'master' version—then you can use the drive for carrying the softcopy of your thesis around. Also consider backing up onto file servers at your institution, and be aware that there are internet services that provide file storage. None of these is absolutely reliable; so why not use all of them?
- Print out your thesis every now and again—the whole thing every 3 or 6 months, or a chapter when it is (temporarily) finished with. This also gives you a hard-copy to scribble on.
- Create explicit versions. When you save your document to disk, you automatically delete the earlier version of it by overwriting with the new version. (Some software is slightly more flexible, and automatically keeps both 'current' and 'previous', but earlier versions are still lost.) For example, if you are reorganizing the structure of a document called 'Ch 08 Discussion' and have saved once or twice, then decide you don't like the new version after all, you will have lost the original. Therefore, if you have any doubts as to whether you might want to keep the earlier version, make a copy of it and call it 'Ch 8 Discussion backup 2011-03-19' before you start editing.
- Create whole directories with complete versions of the thesis. Include the date of the version in the directory name—thesis-2009-03-19 is unambiguously from March 19 2009. A typical thesis is a few megabytes; in 2010 a hundred megabytes of disk costs less than one cent. You can afford to create a version as often as every day. Back these up, too.
- Use logical names for the thesis components. A student of mine, Jim, gradually got into difficulties because his naming scheme broke down. He kept every version, but they would have names such as 'results-off', 'results-no-refs', 'results-keep', and 'results-valsNotChecked'. He was adding results he thought were interesting, but not to be used in the final thesis, to results he wanted to keep. While 'results-off' had been used for official runs (that is, data he was happy

---

[4] However, it is not a good idea to open up the files while they are on the USB drive—instead, first copy them to your hard disk. Currently (2014), the USB drives can burn out if they are being accessed continuously. If used as backup only, though, even the cheap ones will last for years.

with), they weren't necessarily as up to date as 'results-valsNotChecked'. He soon lost track of what was where.

- Make sure you have a single location, maybe that USB drive, where you keep the master version. *Always* make sure you are editing the most recent version.

Whenever you prepare a document for your supervisor, such as a review on a topic or a draft chapter, label it in the header or the footer with the following: page number; the name of the document (and give it a name that clearly identifies it, so that your supervisor doesn't get a document labelled 'Draft Chap. 3' and not know whose Draft Chap. 3 it is); and the date, so that you and your supervisor don't get tangled up as to which version of Chap. 3 is which. If your supervisor agrees, put a due date on it—the proposed day on which your supervisor will return it to you.

Writing Style

Most books on writing theses deal with the art of writing and presentation. They usually deal also with the conventions that support good expression, namely grammar and punctuation. This is not a book about writing style, but rather about structure and coherence, and for advice on good grammar and so on you should look elsewhere. Likewise, I cannot deal here with all the errors that I have come across in my reading of draft theses. I recommend that you buy at least two writing books: a style manual and a guide to good writing. Read them thoroughly, and keep them on your desktop. Such books will tell you all you need to know; more than you can take in at first.

We develop a writing style long before we start to write a thesis. Some people can effortlessly write beautiful, clear, direct English that aids communication. Others have writing styles that hinder the reader: verbose, ungrammatical, turgid, laboured. The strange thing, I've noted, is that some of the worst writers seem to be unaware of their faults, and have no desire to improve. It would take another whole book to deal with this. I suggest that you ask two or three people you respect to do you the favour of telling you what they think of your style—indeed, your supervisor may annoy you by doing this without being asked. When people criticize your writing, don't be defensive. Instead, thank them, and set about to improve your work.

Not only do some students not see their own faults, but they don't see faults as a problem, that is, they do not appreciate the importance of acquiring a style and writing to a good standard. One of my students, Liz, wrote so sloppily that it undermined the value of the work, with sentences that didn't parse, inconsistent notation and nomenclature, and even basic faults such as many misspelt words. There was a strong mismatch of expectations between Liz and me—she saw no benefit in writing well. She sometimes invested more energy in disagreeing, or in deflecting the argument by pointing at ways in which her work was strong (and it *was* strong), than it would have taken to correct the writing. Of course it is possible for a supervisor's expectations to be unreasonable, but in this case the problems were significant, and her reluctance to acknowledge faults meant that she could not develop as a researcher.

There is an important point here that many students seem to miss. Writing well is not just about adhering to an arbitrary set of rules just for the sake of it; it is about the messages you send to your readers. Writing that is full of mistakes says that the author is lazy or incompetent; lazy, incompetent people don't do solid research. Writing that is impenetrably complex or knotted up says that the author is incapable of clear thinking. It is these kinds of messages that you are trying to avoid. And remember too that good researchers are busy: if you want other people to read and appreciate your work, you have to make it easy for them to do so.

Liz was a native English speaker; in contrast, Marsha, who came from central Europe to undertake graduate study in Australia, wrote badly but felt otherwise. She would react to criticism by searching for material in my papers, or in other papers written in my research group, that she felt had the problems her work had been criticized for. Feeling stressed by the pressure to complete her PhD, she would seek reasons to resist undertaking additional work, some of which were not rational. For example, she felt that students she disliked must be less competent than her, and would have difficulty accepting criticism if it wasn't also made about the students she believed herself to be in competition with. The moral is that one should listen to critics, and assume that others aren't trying to undermine you but genuinely want to help you to do your work better.

Thesiese

A particular style fault, namely *thesiese*, seems to afflict some students. Such students have become psychologically oppressed by the problem of impressing the mythical examiners whom, they feel, will respond best to a particular form of language. It is easier to recognize the presence of thesiese than to define it. Here are two examples I've encountered:

> The assessment will require an analysis and application to the study area of available knowledge about human practices and landscape and weather scenarios influencing fire behaviour and occurrence.

I think this means something like: 'To assess [to assess what?] we will need to know how landscape characteristics, weather conditions and human practices in the study area contribute to the outbreak of fires and influence their behaviour once they have started'. Consider this second example:

> Implementation targets must be firmly established and the market and political institutional impediments identified and rigorously addressed if meaningful progress is to be made.

I don't know what this one means. The more I try to rephrase it, the less sense it makes.

Writers of thesiese often use the passive voice ('targets must be firmly established ... and impediments identified'; if the active voice had been used instead, it would be clear who had to establish the targets and identify the impediments); their verbs are activated by other verbs ('will require an analysis and application', rather than 'will analyze and apply'); their sentences are long and complicated; they prefer long and seldom-used words to the short equivalent words common in everyday communication; some phrases carry little information ('scenarios'; 'political institutional impediments'); and so on. You can see from these examples why thesiese does not impress examiners. You are far more likely to impress them by using simple, direct words and sentences. Remember that the university has asked them to look for critical thinking, not obfuscation. Things to avoid:

- 'Carpet-bag' sentences. Allow me to illustrate. Such sentences, like this one, which I hate to encounter in a thesis because I know they will be impossible to correct, sometimes seem to arise from lack of confidence, where a writer isn't quite sure what she or he wants to say, or may even have lost track of what they want to say, and so says several things in the one sentence that might almost be contradictory; and sometimes arise from overconfidence, where the writer genuinely has a complex concept to communicate to the reader and tries to discharge the whole explanation in a single sentence, and the effect is the same, namely, a confused mess with excessive, or even absurd, punctuation, and a strangled syntax that no likely reader will be able to digest, if they even get that far. That was a carpet-bag sentence—get the idea? Most examples are not punctuated that carefully, either. If you can cut a sentence into parts without destroying the meaning or tone, do so.
- Excessively long paragraphs. I know there is a culture in some disciplines of showing intellectual power through complex writing structures, but is it a true display of intellectual virtuosity, or mere showing-off? Examiners are not impressed by ego.
- Cliché, homespun phrasing, and folksy metaphors.
- Empty adjectives and phrases. Examples include *very*, *quite*, *accordingly*, *of course*, and *the fact that*. If a word or phrase can be deleted without affecting the meaning of the sentence, then delete it. The sentence will probably look stronger afterwards.
- Pointless qualifiers. Examples include *may*, *might*, *perhaps*, and *possible*. Like empty adjectives, such words can be a kind of padding that the author believes gives the writing a more academic 'feel'. I suspect the underlying motivation is that academics are not supposed to express absolute opinions, as there is always room for doubt, but the overall impact can be that the reader doesn't learn what the writer is trying to say—every assertion is, in effect, qualified by 'maybe, but then again, maybe not'. Only use a qualifier if you really need to.

Use of the Passive Voice

You will write more clearly if you use the active voice for verbs rather than the passive voice. Although it is not always appropriate, active voice should usually be your first choice. Here is an example of passive voice:

> The agricultural reforms *have been seen* to be successful, which has led to a surge in agricultural production and productivity, contributing to higher savings and investment, and the release of large amounts of labour for employment in emerging rural industry, notably town and village enterprises (Wheeler and Still 1992).

Notice that it is not clear who saw that the reforms were successful. Even by the end of the sentence it is not clear whether Wheeler and Still saw that they were successful, or whether they stated that the consequent surge in agricultural production released labour for other activities, or both. Using the active voice forces you to clarify, as in this revision:

> Chou and Yung (1991) showed that these agricultural reforms had led to increases in both agricultural production and productivity. Wheeler and Still (1992) claim that this increased productivity released labour for emerging rural industry, notably town and village enterprises.

Note that using 'showed' rather than 'have been seen' enables me to avoid using the vague word 'successful', because I now define it, and know exactly who did what. Note also that I have made Wheeler and Still the subject of the second sentence, rather than just being a reference at the end of it, so the reader knows who was drawing the conclusion about the effects of increased productivity.

Use of the First Person

As the structures of science—papers, reviews, disciplines, and so on—became elaborated during the nineteenth century, the idea developed that science was impersonal, that a scientist was a disinterested observer of the unfolding of new knowledge. It followed that scientific researchers could not claim any personal credit (or even display excitement) over their discoveries when they came to report them. Theses, reports and scientific papers had to be written in the third person, as if someone else had made the discovery. To remove the presence of the writer from the text, scientists resorted to use of the passive voice.

Originally scientists wrote perfectly clearly in the first person: 'I observed that …' or 'We observed that …' But over time, they began to use the third person: 'The researcher observed that …', or, if this wasn't clear enough, the incorrect—and confusing—'This researcher [which one?] observed that …'; or often the awkward 'The present writer observed that …' Worse, they began to use the passive voice: 'It was observed that …', or, since the use of the passive may prevent us from knowing *who* observed, 'It was observed by the present writer that …' Using eight words where three did the job very well is one of the building blocks of thesiese.

This tendency has never been as bad in the humanities, where authors are allowed to take positions, and the first-person, active voice is permissible and sometimes even encouraged. Nevertheless, writers in the humanities often hide behind the anonymous third person.

When writing your thesis, what should you do? A thesis examiner may belong to the old school. Rima decided to use the first person plural in her thesis: 'We can see that …', meaning, 'I the writer and you the reader can see that …' I did not discourage her, as I thought it came up well. But one of her examiners didn't like it, and grumbled at her use of the 'royal' plural. Perhaps he didn't like being pushed into agreeing with what Rima was saying. Perhaps the use of the first person just bothered him. Fortunately, he was not the kind of examiner to let personal prejudices get in the way of a fair appraisal, and he passed the thesis. Such cases are increasingly uncommon—of around thirty-five theses supervised by the authors of this book in the last decade, all used 'we' and not one has been criticized for it—but do consider the views of your supervisor.

There are some situations where it is just plain silly to stick to the third person. Examples are given below from theses in the social sciences that were dutifully written in the third person, except for the situations listed. They passed.

When you are recounting personal experiences:

> I arrived in Sri Lanka in the first week of January 1991. By the end of the third week I had dispensed with the research plan I had brought with me. Seven days of being in Sri Lanka taught me more about the practical circumstances of research there than the 7 months of reading had previously.

When you are stating personal opinions:

> I would like to stress, however, that I am convinced that what has been written about the Grameen Bank reflects reality in the field.

When you are explaining the choices you made in research procedure:

> I am not a native Bicolano so, though I understand the Bicol dialect, I cannot speak it fluently, and people laugh at my funny accent. Though the average village person can understand Tagalog (my own dialect, spoken by the people in Manila), the old and poorest cannot. Moreover, because of lack of use, I had forgotten most of the Bicol terminologies. Thus the Research Assistant was my interpreter. [Try putting this into the third person!]

However, there are cases where the first person is inappropriate—you should not imply, for example, that someone else's experiences are you own. For example, if your thesis includes material drawn from a co-authored paper and someone else, say Smith, did the fieldwork, the uses of 'we observed' in that paper should be changed to 'Smith observed'. And, sometimes, overuse of the first person can blur the line between the observer and the cause of the events being observed—for example, don't imply that you are responsible for the laws of physics that led to a chemical reaction!

For a definitive perspective, go to the top five journals in your field and determine what style is used. Look, too, at the use of voice to see if it is first person singular, active ('I investigate') or perhaps third person passive ('the event was investigated'). If your work is cross-disciplinary, settle on a single style so that your work is consistent.

Verb Tenses

Some non-English languages don't use tenses, but rely on the context to indicate whether something happened in the past, is happening now, or may happen in the future. In English we have such a rich collection of tenses that we often get them wrong. This creates problems not only for students whose native language is not English, but also for English-speaking students. Here are some general rules for tenses in academic writing:

- Use the past tense when you are reporting what you or others did at particular times: 'Smith and Jones *reported* the results of their investigation of housing trends in their book published in 1985.'
- Use the present tense in an introduction to a chapter or section or a table where you are outlining its contents: '"Why We Have Trouble with New Tasks" *is a review* of the state of the housing industry in the USA after World War II.' 'Table 3 *shows* that in all countries car ownership increases with GDP per capita.' Future tense is incorrect.
- Use the present tense when you are discussing the implications of some work of yours or others: 'Smith and Jones reported the results of their investigation of housing trends in their book published in 1985. This work *reveals* that the poorest group in the community *find* it almost impossible to find adequate housing.' Note that we are shifting from past to present tense in the same paragraph.
- Use the future tense when reporting the implications of your or other work. 'With evidence that discrimination remains common, it *will be* necessary to make changes to the legislation.'

Punctuation

This book is not the place to learn basic grammar, but there are mistakes in and usages of punctuation that seem to be peculiar to academic writing. I briefly review a few of these.

Commas In general writing, commas should be used sparingly. In academic writing, a few more commas may be warranted to avoid ambiguity.

- Between items in a list, and before the final *and, etc.,* and *or*: 'their own surveys, interviews, observations, experiments, and so on'. In a sentence such as 'the four main groupings were children, employees, pensioners and the disabled, and the unemployed', the consistent use of the list comma allows the reader to easily see that 'pensioners and the disabled' are a single grouping.
- After transitional words such as *however, nevertheless, moreover, therefore,* and *similarly.*
- To put a word or phrase in parenthesis. (To test whether something should be in parenthesis, try omitting the commas altogether; the sentence will lose some information, but should still make sense.) One comma must be placed before the word or phrase, and one after it: 'and, as discussed previously, the three plaintiffs were approached later'. Leave out both, or none.

Semicolons The main use is to separate parts of a sentence that are too closely related to be broken into separate sentences: 'Writers of thesiese nearly always use the passive voice; their verbs are activated by other verbs; their sentences are long and complicated; they prefer long and seldom-used words to the short equivalent words common in every-day communication; jargon is rife; and so on.'

Colons The three main uses of the colon all have a sense of *introducing* something that is to follow: a list, or an explanation, or a quotation. 'These systems make checks such as: whether it contains a verb; whether it is overly complex; whether the subject agrees with the verb; or whether stock phrases being used.'

Dashes and hyphens They are different, and each has its own specific uses. You should find out how to create both on your word processor.

- The *dash* (or em rule, '—') has two principal uses: to indicate an abrupt change in the sentence structure, and to indicate material that is in parenthesis. Use sparingly. As with all parentheses, use two or none.
- The *hyphen* '-' is used to build up complex words. The most common are words built up from suffixes such as *sub-* or *non-* (these suffixes should never stand alone as separate words). As time goes on, some of these complex words become words in their own right, and no longer need the hyphen: thus *sub-zero,* but *non-conformist.* Consult your dictionary.

Compound adjectives can be tricky. One student came out with 'sulphur reduced residual fuel oil fired brick kiln'. Where should he have put the hyphens? Another produced a 'non-cost of living indexed pension'. The first of these is a mixture of compound adjectives and compound nouns, some of which don't take hyphens (such as *brick kiln*). The best solution is to break it up a bit. I suggest 'brick kiln fired with sulphur-reduced residual fuel-oil' and 'pension that is not indexed for the cost of living'.

Exclamations Avoid them! They are annoying!

Capitalization Some researchers seem to like capitalizing Important Terms and descriptions of Common Processes, almost as if they were headings embedded in the text. This excess of uppercase letters seems to say 'the author is unfamiliar with academic English'. If the meaning is still clear with a lowercase initial (and the word isn't a proper noun) then don't use a capital.

Brackets Curved brackets (parentheses), and square brackets have quite separate uses. Don't use them interchangeably; and don't use other types of bracket, such as curly brackets (braces), except perhaps in mathematical expressions; they don't have an agreed meaning.

Quotations The principal use of quotation marks is to enclose the exact words of a writer or speaker, whether or not these form a complete sentence or sentences. For this purpose, use single quotation marks everywhere, and double quotation marks only for quotations within quotations—or, if you prefer, the other way round. But be consistent.

There are other ways of indicating quotations, and other uses of quotation marks. These are the main ones:

- Long quotes from the work of others, say longer than thirty words, should not be designated by quotation marks and contained within the normal text, but instead should be presented as a separate block. The whole block should be in slightly smaller type, indented, with space above and below. Quotation marks are not needed, and should not be used. And the quote should not be in italics.
- Quotation marks are used to indicate that the enclosed words are the title of a chapter in a book, a paper in a journal, a poem, and so on.
- Quotation marks were used to indicate colloquial words in formal writing, or technical words in non-technical writing. However, it is now common to use italics for this purpose. (On this point, note that use of underlining is obsolete.) After the first use of the word the quotation marks may be omitted. Many writers extend this use by putting pet words or humorous expressions in quotes. It is best to avoid this as much as possible: it can become a bad habit.

Link words We use link words to indicate the logic flow in a passage of text. They are of two kinds: *conjunctions,* which are used inside sentences to link clauses, and *transitional words,* which are used to link a sentence to the one that preceded it. Many writers seem to use them interchangeably. This is a great source of confusion. Commonly used conjunctions are *but, although, unless, if, as, since, while, when, before, after, where, because, for, whereas, and, or,* and *nor.*

Transitional words are used to link one sentence to the next. Commonly used transitional words are *however, thus, therefore, instead, also, so, moreover, indeed, furthermore, now, nevertheless, likewise, similarly, accordingly, consequently,* and *finally.* We also make use of transitional phrases: *in fact, in spite of, as a result of, for example,* and *for instance.*

The confusion arises because some of the transitional words are commonly misused as conjunctions, as for example 'in such reports the underlying theory used as a framework for the investigation might be reviewed *however* it is unlikely that new or improved theory would be developed'. The opposite fault is also common—conjunctions used as transitional words.

Repeated words In creative writing, or writing for popular publication such as newspaper articles, the usual advice is to avoid repeating words. In academic writing, such avoidance of repetition can be downright annoying—if you have a precise thing you need to say, use the precise word to say it. I recently read a paper that referred to 'the Levukans', 'the locals', 'the inhabitants', 'the natives', 'the residents', 'the subjects' (of the research, not of a monarch), 'the villagers', and others. In the context of this paper, all of these meant the same thing. The problem is arguably more acute in technical writing, where for example the verbal gymnastics used to avoid reusing 'synchrotron' (machine, system, installation, equipment, and so on) made a paper completely unreadable.

A related issue is jargon. Some authors have a bizarre compulsion to completely avoid use of technical words, perhaps in the fear that they will be condemned for using jargon; or they use technical words in all sorts of inappropriate places, perhaps to demonstrate that they know them. Both are examples of thesiese. There's nothing wrong with using a technical word; yes, some people won't understand it, but are these people the experts who will be reading your work? Write for the right audience.

Repair words In spoken language we often use what we might call 'repair' words to patch up sentences that are going wrong. We can get away with a lot in spoken language, because we can hear the tone of voice and the pauses while the speaker is struggling for the right way of saying what he or she means. But in the written word it just results in a mess. Here is a list of the most common repair words: *regard* (as in 'as regards' or 'in regard to'); *terms* (as in 'in terms of'); *aspect*; *issue* or *situation*; *relation* (as in 'in relation to'); *compared with* or *to* ('if we look at elephants we find that they are large compared to lions' rather than 'elephants are larger than lions'); *address, embrace,* and *resolve* (usually 'issues' get these); *relative*; *former* and *latter*; *basically.*

I'm not suggesting that these are not legitimate words, but rather that authors use them to fudge things when they can't really work out what they are trying to say. Take, for example, the sentence, 'One issue that has to be resolved is the issue of housing for low-income people'. The writer is trying to hint that there is some problem with housing for this group, perhaps price, perhaps availability, perhaps the whole political system that makes it nearly impossible for them to get decent houses. Unfortunately, it says nothing clearly, and revision is required. For some phrases such as *in terms of* or *with regard to* reconstruction may be quite simple: reversing the sentence order, or changing from passive voice to active voice. But for that word *issue* you will have to take the time to say what you really want to say. Often you will need two or three sentences to do this. Using the word *issue* is just a way of avoiding the labour of stating clearly what you want to say.

Misused words These are words that have a strict definition, which is then used in a metaphorical sense related to their original meaning. Here are a few of my favourites, but there are many others: *parameter*; *focus*; *scenario*; *viable*; *empowerment*; *situation*;*highlight*; *core*; *explore*; *stem*; *paradigm*; *mainstream*; *significant*; *key*; *ramifications*; *aspect*; *facet*; *huge*; *immense.*

There is nothing wrong with these words, but if you use them metaphorically, do so in a way that is consistent with their original meaning. Take, for example, *highlight.* The dictionary meaning (Webster) is 'the lightest spot or area in a painting'. The metaphorical meaning is 'an event or detail of major significance'. Use it only in this second sense. Another example is *viable.* The dictionary meaning is 'capable of living'. The metaphorical meaning is 'capable of existence and development as an independent unit'. Don't stray beyond this second meaning. When in doubt, look it up in the dictionary.

Appendices

Appendices or annexes, as we can tell from the derivation of the two words, are things appended or tacked on to the main text of a report or thesis. They do not participate in the main thread of argument, but have been included to support it in some way. They might establish the context of an item in the main text, or give the derivation of an equation. They are often used as a repository for raw data. They might give a sample of a completed questionnaire (in this case the main text would describe how the researcher constructed and administered the questionnaire, and would summarize the results obtained).

How do you decide what you should include in the main text, and what you should relegate to appendices? In most universities PhD candidates are given a word limit of 100,000 words, exclusive of appendices. Students often find that they have exceeded this limit, and a typical reaction is, 'Well, I'll have to put something in an appendix then'. Although this sounds a bit arbitrary, it does make sense. The university is saying that if your argument takes more than a hundred thousand words, it is too diffuse, and probably you have included material that you *should* put into appendices. But what should go? The test is simple: any material that would distract the reader from the development of the argument in the main text should not be there, no matter how interesting it is, or how essential that the reader have access to it. An obvious example is the inclusion of detailed references to enable the reader to follow up material quoted from other works. It is essential that references be included in the report or thesis, but quoting the detail of them in the middle of the main text would be quite distracting. A list of references at the end of the report is a type of appendix. However, this is a test for *excluding* material from the main text, not for including it in an appendix. It might be that you should exclude it from your thesis altogether.

We need another test to decide what to *include* in appendices. Robert gave me a draft chapter of his thesis to read, and it was obvious to me that, although he had put some of the material in an appendix, much of what he had left in the main text failed the first test: it interrupted the flow of his argument. I sent him off to apply this test for himself. In his revised version, with the superfluous material relegated to an appendix, the argument in the text flowed nicely. But to my astonishment I found that one of the appendices itself had an appendix—the original appendix was now tacked on as an appendix to material that was itself now relegated to an appendix. He had written it, and couldn't let it go. Finally, perhaps to humour me, he omitted it altogether. Don't include material in appendices unless you are fairly sure that it is necessary to support your argument. If it is a thesis, try to imagine yourself in the examiner's place and ask, 'Would I want to follow this up?' This is not a strong test, but it is worth applying.

As appendices are there to support material in the main text, you should insert a reference to them at the appropriate point. It is pointless to include appendices that you don't refer to in the text. (You may think that this is too obvious to mention, but I have often seen stand-alone appendices.) Give the appendix an appropriate title— not just 'Appendix 3', but 'Appendix 3: Derivation of the logistic equation'—and briefly explain its purpose.

Plagiarism and Research Integrity

Over the last couple of decades, one of the changes to academic culture has been the development of national and international codes of ethics. Correspondingly, most universities have adopted guidelines for the conduct of research, and funding bodies have policies reflecting their perspectives on such guidelines. These cover general issues such as conflict of interest, authorship, fraud, expenditure of grant monies, rights and responsibilities, data handling and preservation, consent of experimental subjects, and plagiarism. There are also discipline-specific guidelines that dictate, for example, how data may be used or who has access to it. It is the responsibility of all researchers, including students, to be aware of the policies and guidelines that apply to their discipline and institution.

Most of these issues concern the conduct of research, but one, plagiarism, concerns how it is presented. Plagiarism is a fundamental issue of academic honesty and instances of it can provoke strong responses. An underlying cause is that academics' reputations are based on what they have written, so that when one person reuses another's text it is perceived as a particularly threatening form of theft. It also has a separate significance in the context of a thesis, because an examiner's judgment of a student's work is not just based on what they have done but also on how the work is presented.

I'm startled when a student expresses a desire, as Dohka did to me recently, to reuse the background chapter from the work of so-and-so, because, in his view, it so perfectly encompassed the background he needed to present. Such remarks may be made in innocence—or rather, ignorance—but they betray two important failures of thinking. One failure is the idea that someone else's background could be perfect. No piece of research, or writing about research, is final or complete. Even if so-and-so was working on the exact same topic, it is unlikely that it would be impossible to improve their work. Also, it is hard to see how it could be perfect for a new project; surely Dohka was not working on a problem that had already been addressed, and surely new work had appeared in the meanwhile. I am pretty confident that, had Dohka carefully analyzed the chapter he wanted to copy, he soon would have found many things he would have done differently.

The second failure is the lack of understanding of the examiner's point of view. The examiner wants to understand whether the student is a rounded researcher, capable of appreciating and interpreting the work of others as well as of undertaking new work. If the background is someone else's work, how can the student be assessed? In this context it is important to remember that a background chapter can be, in its own way, an original contribution, because it may offer a new understanding or synthesis that by itself helps to advance the field.

Plagiarism may of course consist of much less than the reuse of an entire chapter; one or two hundred words of direct copying would usually be regarded as a breach of ethical standards sufficient to trigger some form of disciplinary proceedings. Nor does it matter whether the original material was an academic paper or, say, a web page; it would still be regarded as copied. Moreover, considering that academics do a lot of reading, and that your thesis will be examined by one of only a small number of true experts in your field, it is surprisingly easy to get caught. Developments in search technology have made the detection of copied material very easy and accessible. Unless clearly attributed and put into quote marks, all of the text in your thesis should be your own.

However, the text might not necessarily be new. You will have generated some text for other purposes, in particular for papers completed during the course of your thesis. Your university probably has some guidelines on how to acknowledge this work; you may, for example, be required to include in your thesis a list of such papers and to identify the pages or chapters where the papers are incorporated. The main difficulty that arises is when the papers were co-authored. If you wrote papers with your supervisor, then it is generally felt to be legitimate to use them if the presentation was principally your work—a paper written by your supervisor around your results should not be used (though you can use the results). If you wrote papers with a fellow research student, it may be that only one of you can report it; check your institutional guidelines. If you do strongly feel the need to use material from another work, you have two choices. You can explicitly quote it—but not excessively, because the bulk of the text should be your own—or you can paraphrase it. For the latter, the simple exercise of reading the work you wish to paraphrase, making *brief* notes, then writing from these notes (preferably some days later) can avoid any risk of plagiarism. In such cases you should still make clear that your writing is based on the work of someone else, and give abundant citations; use of someone else's ideas or thoughts without due credit is another form of plagiarism.

In a related question, how much should you ask—or even pay for—someone to edit your own writing? The short answer is 'nothing': ideally, each word in your thesis should be your own, and yours alone. However, it is accepted practice at university learning support units to provide assistance during a consultation session with you for a single chapter. In such sessions, they will ask you to reflect on what you have written, ask you if you can self-identify tangled prose and faulty grammar, and suggest to you some strategies and materials to improve. University learning consultants are asked to not edit your work, rewrite material, or fix up poor structure.

Summary of Chapter 3: Mechanics of Writing

Learn how your word processor supports authoring of long documents:

- Writing and word-processing habits learnt on short documents such as essays are not effective for thesis writing.
- Use the right word processor for your academic community.
- Use referencing tools for maintenance of chapter numbers, citations, and so on. Be aware of the distinction between presentation and content. Develop a style template based on the most common style of your field.
- Use tools to check spelling and grammar, but check manually as well.
- Use appropriate drawing and graphing tools, which may not be the ones you are familiar with from other tasks. Make sure the results look professional.

Document management:

- Develop a systematic method for determining what constitutes the 'master' copy of your document.
- Back-up your work frequently and in a variety of ways.

- Explicitly maintain versions, with dates. Don't rely on 'track changes'.
- Always label draft documents with the document name, your name, page numbers and the date. Start the document with a table of contents.
- Be systematic in your maintenance of a database of references.

Writing tips:

- Get feedback on your style early in the thesis-writing process. And listen to it.
- Write in a natural, conversational style. Be easy to read rather than self-consciously intellectual. Avoid the passive voice.
- Use an appropriate referencing style.
- Be alert to common problems in academic writing.
- Do not plagiarize or in any way inappropriately use text that is not yours. Do your own writing.

Online resources:

- There are plenty of online guides to preparing theses with specific word-processing tools; a web search such as 'using Word to write thesis' or 'templates for thesis' will return many useful results.

Chapter 4
Making a Strong Start

In the first months of a typical PhD (or, to a lesser extent, the first weeks of a minor thesis), you need to get into the habit of thinking and working like a research student. Your supervisor may set you some reading and introduce you to what your predecessors have done and to the complexities of your chosen field. In a technical discipline, you might choose one of these papers and identify how you could attempt to produce similar results; in a history project, say, you might start exploring what primary sources are available. To consolidate this reading, and to ensure that you understand it with sufficient depth, you may be asked to write a review showing how the field has been developing and what the current challenges and problems are.

That is, your first piece of research will probably be an initial review of existing work in the area, or perhaps an assessment of the state of the art in some technical domain, or an exploration to identify the dozen or so key papers and researchers that will initially influence your work. You may be tempted to write this as a 'brain dump', a compendium of abstracts of all the papers you have read, but such an approach requires little intellectual effort and misses the point. For example, finding some relevant papers may only take a few minutes with a web search engine, and does nothing to develop your critical understanding of the field.

A more constructive approach is to write a chronology of how the topic has developed, or, better still, an encyclopedic review, using as a structure one that had been used previously by an eminent worker in the field. Writing such a review involves gathering papers under headings and discussing similar results together in a 'compare and contrast' style. This initial step will become an ongoing process in your research: discovery and critical analysis of related work is part of the typical weekly routine of successful students.

As this initial reviewing work develops you should be able to define your topic more carefully, and put some limits around it. You will gradually find out what the real unanswered questions are, either through your reading and analysis or, in an empirical discipline, perhaps through experimental work that evaluates current approaches. You may be able to reformulate these as propositions or hypotheses. And you might be able to begin to devise methods for answering your questions or testing your hypotheses.

D. Evans et al., *How to Write a Better Thesis*, DOI 10.1007/978-3-319-04286-2_4,
© Springer International Publishing Switzerland 2014

Without realizing it, you have not only started your research but you have started writing your thesis, and could even begin to reshape some of those earlier pieces your supervisor asked you to write into thesis-style chapters. When research and writing proceed simultaneously, there are three potential benefits. I have already considered the first: arguing out your ideas in writing helps you to think more constructively about them and helps you to identify the processes that enabled you to reach these insights. All of this should lead to better research questions or hypotheses, and better design of your research program. The second benefit is that, if you start to write at an early stage, you will be well into your writing before you have seriously commenced your own analysis or experiments. Therefore, you don't have the formidable task of 'getting started' on your writing when you have all but finished your research, because you will have started much earlier, and have been getting valuable feedback on your ideas and writing throughout your candidature. The third benefit is that it helps you to give shape to your project, including the thesis that reports on it, at an early stage. To explain this, I now outline how you might proceed.

Creating a Structure

Earlier I described a standard structure for a thesis. Perhaps surprisingly, you could devise this structure at a very early stage of the work. To do this, first write a draft of your introductory chapter—the problem statement, the aim and scope, and the steps you think you might take to achieve the aim. You may not feel too confident about writing this introduction, because you suspect that it will have to be modified later, as you get into your work. In this you are almost certainly correct, but that should not prevent you from writing a draft or sketch introduction. What you are trying to do is get started. This sketch might be flowing text, or might even just be a series of bullet points that capture the content you think is important. A good source of inspiration at this stage is to find ten or so theses in your broad area and have a careful look at their contents pages. Some will be good, others poor; analyzing them will help to shape the structure of your own work.

These first steps may be part of the processes at your university. For example, in many institutions PhD students are first admitted to probationary candidature. At the end of an initial period, of say 9 or 12 months, they are asked to prepare a confirmation or progress report, which is used by the university to review whether the candidate has a viable project and appears to be on the right track for a PhD. (The period might be greater for degrees where students spend their first year or more undertaking preparatory coursework.) As noted above, such research planning processes can be seen as the first stages of preparing the final thesis.

These processes are designed to help avoid issues such as students failing to form a clear hypothesis. Burdensome as these processes can seem, they encourage students to define and design their work from an early stage, and I know of many cases where they have helped a student to get a clearer understanding of what they are trying to achieve. Such reports often end up forming the basis of both the introduction and the beginning of theses' cores, as can be seen in the following table.

Example confirmation report	Draft thesis
Thesis title	Preliminary pages
Description of research project	
Introduction to the problem	Chapter 1 Introduction
Overview of relevant research	Chapters 2 & 3 Background chapters
Questions raised by this overview	Chapter 4 Research design
Proposed research procedure	
Research methods	Chapter 4 Research design
Data sources and data collection	Chapter 4 Research design
Research timetable	
Bibliography	

Notice the similarities between the two columns. After the first year, candidates should be able to write a draft introduction chapter, complete with problem statement, aim, and thesis outline; draft background chapters; and a draft 'research design' chapter that sets out the contribution of the thesis. There is also an expectation that candidates have read enough to produce a reasonable bibliography and that they can say how they are going to complete their research—that is, have a timeline for writing and submitting a thesis in the next 2 or 3 years. If you don't get this far in the first year of a PhD, my experience is that you will struggle to finish on time.

As an example, suppose you decide to do research on diet in wealthy countries. Your supervisor, no doubt, will ask what the aim of your research is, but initially you have trouble with this question, as you became interested in diet only because you had noticed that a great deal of food is wasted at home and in shops, and had wondered if this could be changed. However, as a way of proceeding, you settle on the following tentative aim for your project: 'To establish whether consumer incentives can help to improve food utilization'. Starting to write will help you to clarify this aim, and eventually help you define the limits of the study.

You are not yet ready to write a critical review of the literature on food wastage at this early stage. But what you can do is review the appropriate background. There are some things that you will need to be familiar with to achieve your aim, such as: economics of food production; mechanisms for food storage and transport; relevant legislation; and psychology of food purchase choices. When you have read the literature and written a piece on each of these you will be far better informed about your project, and may have revised your aim. It should indicate how you might limit your study. You will certainly be in a much better position to, say, devise surveys of retailers, and examine the economics of better food storage. The writing will not be wasted, as much of it can be used to help draft your final thesis.

Thinking your way into the project like this will help you to write a tentative structure for the first part of your thesis. Of course, there will be a big blank in the last two chapters, which may just read 'Discussion' and 'Conclusions', but you at least have enough to draw up a tentative table of contents.

In my experience, a strong obstacle for some students at this stage is a fear of making a false start and thus of wasting time. However, almost any concrete work early in your PhD is time well spent: it helps you come to grips with the literature, gain experience with tools, and learn to write as a researcher. Moreover, it can be difficult to establish a clear line of research without writing about it—for example, it may well be that the exercise of creating an initial thesis is how you learn that the topic needs to be changed.

A student of mine, Jacob, drafted his thesis structure with chapters, sections, and subsections. He used this structure wisely: whenever something occurred to him, he would add a new section and a few lines of text to describe what he thought he should say. As a result, he had a substantial thesis structure, with ten or twenty thousand words written, a year before submission. But during the final write-up of his thesis the sections got out of control. The structure got too deep—in some places, his subsections had sub-subsections and the subsubsections had titled paragraphs with numbered subparagraphs. He struggled to find headings for many of the elements and had sequences such as 'First Stage ... Second Stage ... Third Stage ... Thoughts' or 'Brundt's Work ... Kavli's Work ... Refinements ... Other Work', each describing just a paragraph or two; he sometimes had a separate heading, completely unnecessarily, for every reference he discussed. Some sections had become irrelevant as the work developed, but he was reluctant to delete them, and so would write a brief, content-free paragraph to justify their existence. In one part of his thesis, he had planned to write a historical survey. During his candidature an excellent review paper was published, saving him a great deal of work, and he could restrict himself to a two-page summary noting key developments. However, he still kept his original headings, despite the fact that each now had just a single sentence underneath. Jacob had failed to realize that the headings that had so successfully provided a guide to his thinking were often unhelpful in the final thesis.

Initial Efforts

Once you have a structure, your next task is to decide which chapter or section you could tackle first, and start writing. Many students find it easiest to start with a factual or concrete chapter, rather than attempt to write material that requires careful argument or complex interpretations or judgments. In a technical area, for example, this chapter might be a description of an experimental design, or a presentation of a series of initial results.

It is not a good idea to start with a 'critical review' of existing theory, as used to be fashionable, indeed compulsory. How one could be 'critical' in these circumstances is quite beyond me—insightful critical analysis requires expertise, and thus such a review is probably the hardest chapter for an inexperienced student to write. Likewise, while you might sketch a background chapter and even write some parts of it in detail, the final content will be strongly influenced by what you have learnt during the course of your research, and thus it cannot be completed at an early stage.

Another model, which works well when you have a clear thesis structure right from the start, involves writing fragments from around the thesis—the kind of writing that, as discussed earlier, is straightforward with a word processor, but was near-impossible previously. In a thesis on, for example, computational modelling of how exotic sugars affect cellular metabolism, you can simultaneously be working on the parts of the background concerning metabolism, documenting the experimental environment, explaining how the 'real world' data is obtained from cells in vitro, and exploring ways of presenting your initial experimental results. That way, if one task is stalled—or is leading to excessive procrastination—you can switch to another and continue to be productive.

As you write you should try to follow the 'rational' structure you have predetermined. But once you start writing to this structure you should let your fingers do the talking: slip uninhibitedly into creative-writing mode. If it is not a complete chapter, write a few notes on the missing sections to indicate how you envisage that they will be organized, and how the material you have written fits in. Print it out occasionally, so that you have a hardcopy to make notes on; put it in a folder with any printouts of any other chapters you have drafted. Every time you go to a meeting with your supervisor or supervisors, take the folder with you. *It is the latest draft of your thesis.*

I notice that some of my students add notes to their thesis more or less every day. When a thought occurs to them, they open up the file and type it in. They use the manuscript draft, right from the first months of their PhD study (or earlier if they are doing a minor thesis), to drive their activity. The draft contains notes on lines of investigation to follow up, bullet point lists of papers to discuss, outlines of discussions and arguments to be filled in later, and so on. A successful strategy is to occasionally browse through this sketchy manuscript, checking the organization—for example, that particular issues are noted in the right place—and, occasionally, taking an hour or two to turn some of the notes into polished text. Over time, it gradually becomes more complete and takes on the appearance of a proper, rounded thesis.

As a result of your own work and thinking, and your discussions on the progress of the project with your supervisors and others, you will no doubt see many gaps and inconsistencies in your draft structure. Revise the structure to deal with these problems whenever necessary. You will retain most of the material you wrote previously, although you might put it in different chapters. For the time being, put any material for which you cannot find a home in an appendix at the back (you will probably reject it eventually, but some of it may be useful). As results start to emerge you will begin to write about what they signify, and drafts of the 'Discussion' and 'Conclusions' chapters will start to emerge. One day you will go to a meeting with your supervisors and realize that the report has taken its final shape—all it needs is a last working through to turn it into the first draft of the complete work.

Writing up at the End

In Chap. 2, I mentioned Karen's problem in starting to write her thesis. When she was a student, word processors and working online were not yet widespread, and the integrated research-writing process that I have discussed was not possible. Instead, we had the 'research first and write later' model of thesis writing.

Despite the flexibility offered by the word processor, it seems that old habits die hard among supervisors. Judging from the seminars I've held on 'Starting Your Thesis', some supervisors are still advocating the old method, and may, in the sciences, even discourage their students from writing until they have finished their main experiments. This brings two risks. One is that the writing is likely to be less polished—we all like to put off tasks we are unfamiliar with, or less good at, and for students with an experimental background writing is often in this category. But, looked at rationally, if writing is not your strength then you should put more effort into it, not less. The second risk is that your thesis will be less of a narrative; you are in danger of writing your thesis around the results of these experiments, and neglecting to tell the reader of what was involved in obtaining these results, not to mention the thinking that led to the experiments. As an examiner, I am never impressed by a thesis that gives an account of ingenious experiments and the results obtained from them, but fails to tell me why the experiments were carried out or what the implications of the findings were. Narrative is essential, and the best way I know to develop a good narrative is to start writing early.

I did manage to get Karen going eventually. I asked her how many important ideas she had identified in those hundreds of papers, ideas that had taken the theory of her topic along another step. To my surprise, she replied almost immediately that there were only four, and explained them. I saw that somehow her unconscious mind had been working on those papers, sorting them, organizing them, ordering them. These four ideas provided the titles of four sections of her review of existing theory, and I suggested she start writing on this basis. She came back in 2 weeks with a brilliant review of the theory that scarcely needed any further change.

My advice to Karen enabled her to get started because it gave her a structure to write to and enabled her to resolve the tension between her rational, conscious thinking and her creative, unconscious thinking. To get started on the thesis, you must acknowledge and *harness* this tension. To do this you must first devise a logical structure for the whole thesis; a good way to ensure that the structure is sound is to write the introductions to each chapter, then string them together to see whether the report develops logically.

Then start to write. I suggest that you use writing to drive the reading. List the papers you haven't read in the place where you expect to discuss them. Your background should mutate from a description of intended reading into an analysis of what you have read.

For the early-start model of thesis writing I suggested that students begin by writing the introductory chapter that described why they were doing the research, what their aim and scope were, and how they intended to achieve their aim. This advice holds for the late-start model also. If you are having trouble with the introduction, you most likely have not yet got the aim of the whole project right (yes, this can happen even after you have completed all your experimental work!), and this is the chapter where polish—the result of repeated revision—is the most important. Put this chapter aside for the moment, and start with another. As mentioned above, a good idea is to start with a factual chapter, such as one describing the study area, or the rationale for selecting a research method, or the design of experiments or questionnaires. These are chapters in which the rational side of writing will predominate, and the creative side will not provide a stumbling block.

However, there is, in the late-start case, an argument for tackling the background chapters first. They are often the hardest to write, particularly in the sciences; the writing involves the chore of carefully reading a couple of hundred papers, and you will need to make hundreds of judgments on their individual contributions and shortcomings. But this means that the background should not be rushed or completed at the last minute—it is also the material you are likely to return to and revise the most often. What's more, with a good draft of the background in hand, completing the other chapters is likely to feel relatively straightforward.

Starting, and Starting Again

Some students seem to have a problem that is the opposite of Karen's. They start an introduction, or some other chapter, look at it, and then start writing it again. They get caught in a seemingly endless cycle of trying to create a good draft, each time thinking that they must get this right before they can do anything else.

Martin was one such student: he came to my office with three different starts. After writing a few pages of an introduction, he told me that it wouldn't work. I gave him some advice about improving it, and he went away to write another. In this second version he had saved one section from his problem statement, omitted some and expanded others. It still didn't work. This happened three times. With each rewrite, he emphasized different points that suggested different areas of investigations. His co-supervisor, like me, would get frustrated: it appeared to her that Martin didn't really know what he wanted to do with his own study. Was he good enough to finish a thesis if he couldn't start? As a defence, Martin said that he was trying to cover all possibilities so that he wouldn't get caught out further down the track. He wanted to be certain that he would pass without any major hold-ups.

It is crucial at this point that you try to make yourself stick to something. This is particularly true if you are undertaking a minor thesis: you have a limited word count and a tight deadline to meet. I knew that Martin had read a great deal of literature, and so I asked him to list the key areas he was interested in. Within a few minutes, he had listed five topic areas. Next, I asked him to circle the ones that he was particularly interested in. Now, he circled three areas. We then looked at how each of these three topics related to what he had already written. We looked at the stated aim in his most recent introduction and then quickly sketched out the structure for each of the three topic areas. By doing this, Martin could see which topic could be reasonably sustained in the time allotted for his Masters degree, and that he would not be able to tackle all five topics or even just the three he was interested in doing. He had to focus and discipline his thinking, and understand the constraints of the type of thesis he was trying to complete.[1]

In a research project for a higher degree it is entirely normal to revise an aim, narrow or expand the scope of the study, or discover that you may need fewer chapters than you initially envisaged. As you work your thinking becomes more sophisticated and you gain confidence and authority in your area. Because of this, I suggest that you look over your introductory chapter on a regular basis. Change a word here or there, check and refine the statement of the aim, add or delete limitations to the study, and think about the overall structure. By doing this, you are likely to keep track of how you yourself are changing your thinking along the way, as opposed to being surprised that you have 'moved on' from your original somewhat naïve perspective. When your examiners read your thesis in a single sitting they will be impressed that the study has remained focused and has not drifted from start to conclusion.

Without focus, it is likely that you will never reach a critical appraisal of the literature. Your treatment of the background material will appear haphazard as it swerves from issue to issue in an attempt to cover a lot of ground. Examiners look for depth of understanding, not a superficial grasp of a wide range of topics.

The Creative Process

Once you start the writing itself, allow the creative side of your brain to work through the argument for you. When you have finished writing for the day, save what you have written. Then go to bed, and sleep on it (the very existence of this expression is evidence that our unconscious thought processes keep working even when all rational thinking has been switched off). Your first task the next day is to look at the chapter outline, then read the chapter as it stands. As well as picking up typographical and grammatical errors you will readily see the results of the tension between creative writing and rational structure. If you find that they are at odds, one of them must give. Either your original rational structure was wrong, in which case

---

[1] A research fellow who worked for me, who already had a PhD, had a similar problem. He would start work on a tightly defined problem, working to a particular paper submission deadline, and would quickly achieve his initial goals. However, in the process he would identify new issues, and set out to resolve them to make sure the paper was 'complete'. Pretty soon the scale of the work would expand to a PhD, or more, and he would have so many doubts in his original, strong result that he couldn't publish. That is, the moment he achieved something—no matter how innovative or useful or surprising—he would race on to supersede or invalidate it, and as a result never had anything he was comfortable reporting. I have no idea how his PhD supervisor managed to get him to submit a thesis! In 2 years working for me, he produced a massive amount of work but no papers. Working to constraints is a key element in being an effective researcher.

you must alter it, or your creative thinking has taken you on a wild-goose chase, in which case you must cut the irrelevant material out. But you now have both creative and logical inputs, and can sit in judgment on the outcome.

A common form of writer's block is the tendency to attempt to write a sentence but to endlessly be critical of the word choices and sentence structure, and to edit it over and over without moving on; and the longer you spend on one sentence, the more completely you lose the thread of what you were trying to say. This is a direct example of the creative–critical tension. That is, writer's block is a particular kind of failure of the creative process.

My student Theo had a severe case of writer's block. He would sit at his desk and type a single word, stare at it for a while, then type another, and then maybe delete the whole sentence and begin again. We tried several strategies, including writing first drafts on paper and writing in tiny sentences: 'We first examined tokens. The set of tokens is determined by the text being examined. Parsing identifies the tokens in the text. A minimal structure can then be constructed.' Yes, this was terrible writing—but at least it gave him words on the page that he could edit into more mature text later on. However, the core problem turned out to be deeper. He had been a part-time student and had lived with his topic for over 6 years, and the ideas had become so obvious to him that he felt his work was trivial (it wasn't) and that he nothing to say. Theo eventually hit on a strategy that worked: he found a colleague, a PhD student who was unfamiliar with his area, who agreed to meet every day for a couple of weeks to listen to Theo explain his work. This process showed to Theo just how interesting and complex his contributions were. Theo wrote down these explanations, giving him the basis of a couple of chapters, and he gradually gained confidence in writing.

Joanne had another form of block. She simply couldn't leave a topic until she felt that she had completed understood it, even when it was only of marginal importance—in some cases, this was rather like refusing to ride in a car because of a lack of understanding of engine mechanics and tyre chemistry. A particularly frustrating instance was her desire to learn advanced statistics (an area she knew nothing about) because of the analysis in a paper she cited. As a result her thesis stalled while she tried to grasp difficult concepts that were well beyond her expertise, and mine too, for no obvious gain—in this case, she could reasonably take the analysis on trust. However, I gradually realized that to some extent this behaviour was because she didn't like writing, and her exploration of side topics gave her a reason to avoid it. When I confronted her, she managed to return to a more focused style of working.

In my own PhD, for many months I agonized over every sentence, struggling to produce even a single brief paragraph in a 2 or 3 hour session. (At that time I worked with pen and paper.) I then found one day, almost by chance, that I had suddenly been inspired and had written several pages in less than an hour (in fact this had happened several times, but I hadn't reflected on it). I realized that the difference had been that I had been keen to write about a particular observation, and had simply jotted down the preliminary material as quickly as I could get my hand to move across the page. But this preliminary jotting ended up in my thesis with almost no edits at all—and it flowed much better than my usual 'agony' text. From then on I made a conscious effort to write carelessly, as I described it to myself, and it usually worked. Sometimes the words wouldn't flow and sometimes I had to edit the text a great deal later on, but mostly I was able to write 10 or 20 times faster than before. With experience, I usually write like this now.

Over the years I have shifted from writing on paper to writing almost exclusively at the computer. However, a few years ago I was struggling with some writing of a kind that was new to me (an essay on an aspect of teaching practice) and had been blocked for days. In desperation I decided to move away from the computer and sit in a favourite spot, away from all the usual distractions of phones and email, to write in a notebook. I had only intended to write bullet points, in an attempt to try and capture the topic of the essay, but within an hour I was writing flowing text, and the thoughts that had been building up in my mind over the previous days were soon on paper. Ever since, I have often kicked off the writing process in a similar way, by brainstorming thoughts into a notebook I carry round with me. Sitting in dull seminars seems to be a fruitful way for new papers to get started.

Writing an Individual Chapter

How do you turn the literature reviews or descriptions of research procedures that you wrote in your first year into thesis chapters? You wrote them as stand-alone pieces, before you were an expert in your subject.

Just as the thesis itself must be properly structured to ensure that the reader always knows exactly what is going on, so must individual chapters. Why is this particular chapter there? What is its function in the thesis? You must make this absolutely clear. The best way to ensure clarity is to write a formal introduction to every chapter. Follow with the business of the chapter itself, then a formal conclusion. Some supervisors consider this to be a rather stilted approach, and skilful writers can get away without formal introductions and conclusions. This doesn't mean that they can do away with them altogether; rather, that they do it less formally and obviously. Most of us don't have such writing skills, however, and I recommend that you write formal introductions and conclusions to all chapters.

By a formal introduction, I mean a piece of text that is designed to explain the role the chapter plays in the thesis. A typical such introduction is organized as about three paragraphs.

- First paragraph: creates a link back to the earlier parts of the thesis, in particular the previous chapter, to make it obvious why the chapter is needed.
- Second paragraph: states the aim of the chapter, what the reader learns from it and how it advances the overall goal.
- Third paragraph: outlines how you intend to achieve this aim. This paragraph often has the 'overview of contents' flavour that so many writers think constitutes an introduction. But it is only one part of the introduction, and without the other two parts the reader struggles for a sense of direction. (Incidentally, writers sometimes literally give it as a table of contents. This is far from helpful. The reader needs to know not only what you will be dealing with in the chapter, but also the logical connection between the various sections.)

Here's an example of a chapter introduction.

From 'Chapter Five: The Place of FC in the Academic EFL Instructors' Practice' in Al-Asmari, Abdul (2008). *Integration of Foreign Culture Into Pre-service EFL Teacher Education: A Case Study of Saudi Arabia*, p. 151. Unpublished PhD thesis, University of Melbourne.

The previous chapter introduced an analysis of the place of foreign culture (FC) in the policy of pre-service Saudi EFL (SEFL) teacher education. As found, the place of FC remained overlooked in policy despite the global and national developments as well as policymakers' positive views about the place of FC in EFL education. Key findings indicate that there is a gap between policy and practice, and the SEFL teacher education was subsequently found to be policy-driven. Considering the contextual and cultural constraints that challenge the place of FC in policy (e.g. centralisation of policy-making), the place of FC may be negatively influenced in practice. Also, such constraints as lack of national standards, resistance of global pressure, and cultural ideologies in SA necessitate developing context-specific concepts for potential integration of FC in pre-service SEFL teacher education.

In this chapter, therefore, I investigate the place of FC in the academic EFL instructors' practice of SEFL teacher education. To do this, I analyse the academic EFL instructors' responses to the interview questions within a focus provided by the previous chapter (See Appendix 3). An important note that I must add is that this investigation is not meant as academic staff appraisal; rather, it intends to examine the place of FC in practice in light of the reported policy conditions and curriculum limitations. Besides confirming findings that emerged at the policy level, my aim is to further shape dimensions for developing concepts and relevant implications to integrate FC within the context-specific conditions.

Structurally, this chapter consists of four thematic sections with relevant sub-sections. In the first section, I discuss the academic EFL instructors' conceptualisation of FC through their perceptions of FC and views about its place in EFL education. Next, I address issues of the place of FC in the actual instructors' practice. In the third section, I lay out a discussion of challenges to the place of FC in practice. Also, I analyse and discuss available opportunities and alternatives for providing exposure to FC in pre-service SEFL teacher education. Finally, a summary of this chapter and focus for the following chapter is provided.

You may be tempted to write far more than these three paragraphs in your introductions, especially in background chapters. For example, the first draft of an introduction to a chapter reviewing approaches to the problem of improving diet in disadvantaged inner-urban communities might include several pages describing the problem, as a prelude to reviews in later sections of the solutions proposed by various schools of thought. The readers will need this review, but if you give it to them before they learn why the chapter is in the thesis and what it is attempting to do you will confuse them. Make it the first section *after* your introduction.

After the introduction comes the main body of the chapter. Its contents and structure will depend on the kind of chapter you are writing and the kind of research you are reporting. Nevertheless, it needs to flow logically from the purpose of the chapter, as stated in its introduction, to its summary of conclusions. This typically involves three or four major sections. Having too many sections in a chapter makes the structure hard for the reader to follow; if you find that you have created a lot of sections, then you need to step back and ask if the structure is still appropriate.

Note that it is bad form to immediately follow a section heading with a subsection heading. This amounts to saying that you have nothing to say to the reader about the section's purpose and contents. Every component of your thesis needs an introduction.

The structure of a chapter should be organic, that is, should be appropriate to the content. My student Surijit struggled with chapter organization. The body of his thesis was designed as a series of chapters with linked experimental results, which were tests of different mechanisms for identifying the source of contamination in stored grains, such as rice that has been bagged for sale in retail outlets. These mechanisms were diverse, ranging from chemical tests on the gases enclosed in the packaging to analysis of the transportation and processing path from field to supermarket. With his supervisor as co-author Surijit had published a strong paper early in his PhD, on a particular chemical test, and had revised this to give one of his first 'contribution' chapters. This established what he saw as the correct framework for presenting research results, including a cost analysis and reflection on sources of error. (He had those precise headings—Cost Analysis, Sources of Error.) So far so good! But he then concluded that all of his chapters had to have the same structure, regardless of whether it was appropriate, and even made curious decisions such as to include a cost formula while announcing that all the variables were unknown and could not be measured, and that the correctness of the formula could not be determined. This shoehorning of all work into the one fixed ordering led to some chapters that were very strange indeed. He needed to let each piece of work take a more natural shape, and to have the confidence to explore what that shape might be.

Every chapter in a thesis should have a conclusion. The reader needs to share with the writer a sense of what has been achieved, what is established now that wasn't established at the beginning of the chapter. And the conclusion should, of course, respond to the stated aim or purpose of the chapter. There may be some exceptions, for example, descriptive chapters outlining information on the characteristics of a study area. Chapters that need strong conclusions are reviews of theory, reviews of available research methods, reports of results, and the discussion (which may be the conclusions to the whole thesis).

I often suggest to my students that they structure the end of a chapter as two sections, a *discussion* or *analysis* and a *summary*. The work reported in a chapter may not yet allow definite conclusions, for example because the case being made in a thesis rests on three separate strands of evidence, each reported separately, and so a formal section of 'conclusions' may not be appropriate. But a typical chapter contains a great deal of detail, and the reader needs the help of the author to sift through this mass: What's important? What overall themes can be identified? What can be observed, or learned? A discussion is also a good place to consider shortcomings or limitations in the work reported in the chapter. The summary then, in effect, replies to the introduction by briefly identifying the chapter's achievements, and sets the scene for the next chapter.

Students often have difficulty with chapter endings. They tend to write lists of what was in the chapter, and fail to state the significance or implications of what was reported. (But don't get ahead of yourself. I have often found that students offer conclusions in background chapters that were informed by insights from their own work. Save these for later in the thesis.)

Below is the conclusion to a chapter of a thesis describing part of a candidate's work.

From 'Chapter Five: The Place of FC in the Academic EFL Instructors' Practice' in Al-Asmari, Abdul (2008). *Integration of Foreign Culture Into Pre-service EFL Teacher Education: A Case Study of Saudi Arabia*, p. 202. Unpublished PhD thesis, University of Melbourne.

As for practice, academic EFL instructors' practice confirmed their static perceptions of FC which was incidentally presented in the form of information about FC. Also, the instructors' practice was negatively affected by their perceptions of culturally sensitive issues, which were policy-influenced and misconceived. Thus, FC in their teaching practice appeared to be largely minimised and disguised. The instructors' generalisation of culturally sensitive issues and influence of curriculum policies encouraged excluding many FC aspects. While previous studies indicated that political factors and community hindered translating teachers' awareness into action (Byram and Risager 1999; Kramsch et al. 1996), political factors in this study are linked to the curriculum policies which were found to provide no support for the place of FC. Further, instructors' background and lack of sufficient cultural knowledge emerged as challenges to the place of FC in pre-service EFL teacher education. For the conservative context of SA, these findings may provide interpretations for a previous study in the SEFL context of Secondary Education where teachers held reservations and worries over the students' beliefs and values (Al-Qahtani 2003).

Confirming findings from the document analysis, there was a deficiency in exposure to FC in the current curriculum worsened by the monocultural context conditions. On the other hand, computer technology emerged as a crucial medium that could provide exposure to FC by offering FC input and facilitate pedagogy and communication. Barriers to the use of computer technology such as instructors' insufficient computer literacy and lack of access were issues to be dealt with as structural limitations at the policy level. However, issues of cultural security online require further involvement on the instructors' part to provide guidance to pre-service EFL teachers on appropriate cultural recourses.

In the chapter that follows, I narrow down the focus of discussion to cover emerging findings of this chapter that can be further elaborated at the EFL learning level. Based on these findings, I analyse the pre-service SEFL teachers' responses and relevant observations to enhance the credibility of assumptions made in relation to the place of FC in practice. Table 42 below provides a concept mapping of the key themes and findings of this chapter and further issues to be investigated in the following chapter.

In the first paragraph, Abdul gives a strong summary of the main findings of his work with academics and relates this to current theory in his field. In the second paragraph, he relates these results to his earlier analysis of curriculum documents and is quite clear about the implications of these findings. The conclusion in the third paragraph provides a strong lead into his next chapter, in which he focuses on the next group of participant data, and finally concludes with a summary table. You might like to check to see whether his conclusions do respond to the aim he stated in his introduction, as given earlier.

Literature

Creating a thesis involves a lot more than writing, of course. For one thing, there's the small matter of finding a question and doing the research! A great many kinds of activities go into creating a thesis. Most of these are beyond the scope of this book, but a critical one is the search for relevant literature.

What constitutes 'relevant' varies between disciplines, but in general the bulk of the citations will be to academic literature, that is, to documents that are accepted by the community as a reliable source of knowledge. These are typically refereed, widely accessible, identifiable, and durable. A PowerPoint slide found on the web fails most of these criteria, as does an email, even if from an eminent authority. Slightly more tricky are primary sources. In the sciences, these might include lab notebooks, say; these are the evidence underpinning your conclusions, but the citable form of these is the discussion of them in your thesis, not the notebooks themselves. In the humanities, primary sources might include, say, anything from wartime diaries to logs of web page accesses; whether they would be cited is discipline-dependent, but most often they would be listed as sources rather than included in the thesis's bibliography.

There are many ways in which we learn beyond the academic literature, including newspaper articles, corridor conversations, resources such as Wikipedia, and the web in general. In most cases it is appropriate to confirm such knowledge in an academic source, and it is the academic source that should be cited. To take a wider perspective, we read for many reasons: to establish that our work is novel and to identify new lines of questioning, for example, as well as to set our work in an academic context. Such reading informs us, but is not necessarily a part of the final thesis. Reading can sometimes be footnoted rather than listed as a formal reference.

To explore the literature, I recommend the strategy of finding the *gateways* then following the *paths*. A gateway is a paper or book or other source you discover through actively exploring, say by using a search facility to try and find new material on a given topic; or you may be introduced to a gateway in a seminar, or by a colleague. In such discovery, don't expect to find what you need by simply issuing a quick query at a search interface. As you search, you will learn new terminology and use it to help discover more references. Also, be broad in your use of tools. General web resources, university portals, and publisher portals all tend to index different materials and may provide very different results.

Gateways lead to paths. Each paper has a list of citations; each paper may also be cited by others, and online resources can help you find these works too. Following these trails, or pathways, is a certain way to discover new literature. Use the web to chase down other work by the same authors and same research groups. Once you've identified other people working the same area, regularly check to see if they have published something new. I also recommend that you plan your search. Identify fields and subfields that you need to explore, and focus on one at a time. And be inclusive in your definition of 'relevance'; for example, a paper can be useful, not just because it is on the right topic, but because it embodies an approach to a similar kind of problem in an interesting way, and it is this approach that you wish to cite.

Don't neglect the traditional source of academic literature: namely, libraries. Students in some disciplines now have little immediate reason to go to the library, since so much of the literature is online, but even in these cases I advise the occasional library visit—and I am often rewarded by students telling me how much unexpected work the visit turned up. Simply browsing the shelves near their usual journals and texts can turn up a great range of new sources.

And don't forget that reading is an ongoing commitment; you should be looking for literature throughout your study. Some of this will be deep reading of critical papers; some will be light reading across an area to get a sense of the current thinking. Some will be a return to search tools to see what new work is being published. Commit to a program of reading, and stick to it.

Having found papers, you need to analyze them, an issue that is examined in Chap. 6.

Styles of Working

Production of a finished thesis involves not just writing in a certain way, creating the right structures, and so on, but also that you work effectively. Very broadly, completing a thesis presents two main challenges. The first is the structure of a research program: several months or years of continuous work on a single project, with only minimal deadlines along the way. In contrast to an undergraduate degree, where each step is tightly specified and completion of a unit means that another fraction of your degree has been achieved, with a thesis you have an 'all or nothing' result based on a single written document. You need to learn to work towards this result from the start.

A key consequence of this line of thinking is to observe that, in the end, only the writing matters. What draws many students into research is excitement about the possibility of making discoveries and creating new knowledge, and in many disciplines this involves undertaking experiments and studies in a lab, or making site visits, or searching for primary sources; the task of writing seems remote. Yet from the point of view of undertaking the work to produce a thesis it is the writing that should be paramount; experiments (and so on) should only be undertaken as required for the thesis. The reality of lab or field work is not like this of course—experiments are used to guide the student's learning, to help form the hypotheses of the thesis, and so on—but it is undoubtedly true that many, many students let experimental work dominate their time and put off the thesis writing for much too long. The fact that writing is a lot less fun than lab work is a large part of this.

The other main challenge is that thesis writing can be a painful grind. Let me rephrase that. Thesis writing *is* a painful grind. Creating a complete first draft of a PhD involves writing up to a thousand words a day of reasonably polished text, every day, for months. Underlying this writing are repetitive tasks of reading papers, maintaining bibliographies, generating graphs, and so on. The text then has to be read, and re-read, and re-read again. It is hardly surprising that many students struggle to stick to a writing routine.

This is one area where every student really is different. Some people write best at night; one of my students habitually wrote from about 10 pm to 3 am. I did much of my thesis writing from 5 am to midday, with a long break for breakfast from say 8 am to 9 am. Some people say they write well in cycles of little bursts, of say 20 min, followed by a half hour of distractions; others find that it takes them an hour or two to settle to the task when they sit at their desk, but that, once settled, they can write fluently. Some work best in silence; some use background music;[2] some like a quiet corner in a cafe; a friend of mine wrote large chunks of a book sitting in a sheltered corner of the stands overlooking the local sports ground. The thing each of these has in common is that the student found a routine that works, and managed to stick to it.

For the main write-up phase of your thesis, a nine-to-five approach is probably not appropriate. You need to be immersed in your work, with a ready knowledge of what bits of your research are discussed where in your thesis, and with a clear grasp of how the work flows. The more time you spend with your work, the easier it will be. This is a contrast to the first two-thirds of your research, PhDs in particular, where the simple fact of the length of the marathon you are engaged in means that you need to balance work and life, and a nine-to-five working style avoids burnout while ensuring that you make steady progress.

I noted earlier the strategy of working on several activities at once, that is, keeping open a range of tasks that can be worked on. This isn't limited to writing; you may well, for example, have a routine that has you at a desk in the morning, doing some writing, and interviewing survey respondents or compiling data in the afternoons. If you have a topic that lets you do the research at a computer, you may be able to fluidly move between a wide range of tasks. In such an environment, though, it is important to be disciplined. The presence of email and the web can be fatal to productivity; when I write, I often physically unplug my computer's network connection. It's unplugged right now.

I ask my students to keep a 'progress log' of how much time they are putting in and what they have accomplished every day, and to be honest with themselves, so that they can use it to recognize when they are not making progress. Everyone has a balance of commitments; it is up to you to get this balance right, and the first step is to recognize that there is a problem. I suggest that you get a notebook and keep it with you. As well as recording progress, jot down thoughts as they come to you, and don't worry about whether they might be silly or irrelevant—this is how you get your thinking started.

As a supervisor, you can't help but be aware that, during their candidature, PhD students lead busy lives. Some marry (or divorce), or have children, or buy homes; or make a substantial commitment to activity outside study. Among my full-time students there have been musicians who are active in bands and orchestras, a sportsman competing in weight-lifting contests, software engineers with their own businesses, and a dedicated volunteer firefighter. Some students exploit the flexibility of a PhD to learn other professional skills, or to study other fields. The best PhD students, after all, have a strong intellectual curiosity and it hardly surprising that their interests are wide-ranging. Nor is it surprising that these interests occasionally become a serious distraction—something that, obviously, should be avoided.

---

[2] Let me emphasize the word 'background'—you are kidding yourself if you think you can be at your most productive while listening to music that consumes your attention, or while watching television for that matter.

These kinds of distraction may be the explanation for a problem that I've seen with a good number of PhD students: a tendency to completely stop work when they hit an obstacle, even when it is trivial. These students will happily work hard at their research, meet weekly with their supervisor and so on, but every now and again will come to a meeting and literally have done nothing for a week. When the student and I explore it, the explanation is that something 'didn't work' or 'didn't make sense'; and for some reason they haven't found something else to do, or found a new way to attack the problem. Worse, while freezing up once would be explicable, such students tend to do so repeatedly, sometimes in several consecutive weeks. In some cases it might be a loss of confidence, but a common factor does seem to be the extent to which the student has a rich life outside study. There is an obvious lesson, about making sure that you are working effectively and so on, but another lesson is that, ultimately, responsibility lies with you and not your supervisor. Problems with effectiveness can stem from anything from difficulties with the research to a personal crisis to simple procrastination to an overload of hobbies and social commitments, and even an engaged supervisor won't know what is going on if you conceal the cause of your difficulties.

Working with Your Supervisor

The process of undertaking a PhD completes a transition from being an undergraduate to being a researcher, a transition that may have started with a minor thesis or some other introduction to research. As an undergraduate, a primary role of the academics you interact with is to assess your skills and award marks. As a researcher, these same academics work with you to share in discovery and the creation of new work. At the start of your PhD, your relationship may be very much that of master and apprentice; in the best cases, by the end it is a meeting of peers.

An exception to this, however, is the write-up process, where your supervisor can guide you but the thesis is the responsibility of the student—the supervisor is not an author, and must arrange examination and so on. Write-up can thus be quite different from earlier student–supervisor interactions such as paper writing that are more of a partnership, a fact that some students struggle to adapt to. During write-up, you also need to be sure to make the most of your supervisor's experience as a reader and referee, and be aware of when these skills are limited; if your supervisor isn't good at grammar correction, for example, you need to seek that skill elsewhere.

I've noticed several styles of relationship with my students. In some cases, we work together on problems as peers and bringing different skills to the research. But this isn't the right relationship for all students or for all research environments; in some cases, our relationship is one of advisor (me) and explorer (student), such as when the work involves a larger step into the unknown. And some students appreciate a supervisor who directs, that is, takes close responsibility for the work and gives a firm structure of activities to work within. This isn't a good working mode for a mature researcher—the sort of person the PhD student wants to become—but can be an effective way of learning.

There is a balance between respecting a supervisor and thinking independently. Supervisors are mature, and are in their role because their knowledge and instincts and experience are reliable; but as your expertise grows, you will sometimes question your supervisor, and some of those times you will be right. Every supervisor has stories of bull-headed students who insist on ignoring their advice and getting it wrong, or, less often, ignoring their advice and getting it right;[3] and, conversely, of students who mindlessly follow their advice and never try to think for themselves. As I said, you need to find a balance.

A colleague, Robert, told me about a student (let's call him Tom) who came to his office and reported the outcome of an obviously silly chemical experiment—Tom should have noticed that the equipment simply wasn't capable of detecting the effect he was looking for. After some discussion Tom suggested that he may have made a mistake. Robert then asked Tom what would happen if the reagents were changed (although they were irrelevant to the failure). A week later, Tom reported back that the experiment still hadn't worked. When Robert asked whether he had expected anything different, Tom suddenly looked troubled; it was clear he hadn't ever stopped to ask whether what he was doing was sensible. Tom asked why Robert had suggested the experiment at all; Robert asked back, 'Why did you do the experiment without thinking?' The lesson for Tom strikes me as a harsh one,[4] but incredibly important: all students must learn the skill of critical thinking, not just every now and again but in every aspect of their research.

Most of the anecdotes in this book concern cases that led to a happy outcome, but not all PhDs go well, and there are lessons to be learnt from the failures. Two cases that are of relevance here are PhDs that went off track due to problems in the student–supervisor working relationship.

Hasrim did not settle into an effective working relationship with me. He initially felt that commencing research would be as straightforward as (if longer-term than) the task of, say, getting ready to teach a new subject: find resources, do some reading, follow a schedule of experiments and investigation, and write up. He had previously completed a Masters thesis at another university, but it quickly developed that this previous experience had not prepared him well, as it had largely consisted of uncritically repeating an earlier investigation undertaken by his supervisor there; he had not even had to search for background literature. (Without the context of having read the supervisor's earlier paper, Hasrim's thesis looks like a sound, independent piece of work, but when they are read together the lack of depth is obvious.) In

---

[3] A student of mine persisted with work on an algorithm that I 'knew' was foolish, and ended up with a strong result and a paper in a top journal. But this case was a rare exception.

[4] This was in the 1980s. In my view Robert should not have deliberately wasted a week of his student's time, although it does seem that the lesson was an effective one. I sometimes use this same anecdote as an example of the kind of treatment of students that was once common but now, happily, seems to be dying out.

particular, he did not seem to grasp the importance of questioning, scepticism, and independence.

Hasrim had difficulty accepting that this previous work was untypical of 'real' research, and to the end of his candidature he was resentful. He was thus reluctant to be fully intellectually engaged in his work, instead behaving more as a research assistant who should refer all questions to me for a decision, even non-issues such as whether to read a paper that was highly regarded in the area. He could become especially frustrated when he felt that I knew the answer to a question but expected him to try to figure it out for himself—which, from my perspective, was a quite reasonable expectation of a student who wishes to mature into a critical, independent thinker! In the end, Hasrim did complete a thesis, but it was relatively weak, and he could have finished at least a year sooner had he accepted that doing research involved him showing initiative (and working in a sustained way).

A fundamental problem was Hasrim's inability to accept that a PhD was not what he had expected, and thus the relationships and responsibilities were not what he had planned for. Many students are in some ways surprised by the kind of work a PhD involves—for example, it can be more grinding, day-to-day, than they anticipate—but doing a PhD can be rewarding in unexpected ways, and a flexible attitude towards your PhD will help you to shape it well and to adapt to problems and opportunities.

It is incredibly important to be honest about progress, or lack of it, most critically with yourself. My student Delia got into serious difficulties late in her PhD because she concealed her lack of progress from me and her other supervisor, for example, by presenting hypothesized results and computer-based simulations—that is, the expected outcomes of experiments—as if they were actual results that she had observed. (Had she kept a careful logbook, this problem might have been detected before she reached a crisis, but such logbooks are not used in every discipline.) Part of the problem was that she was deceiving herself; she 'knew' what the results of an experiment would be and that she was being held up by problems in the apparatus; that is, she thought that the experiment would run properly someday and these results would be observed. In this mindset, we, her supervisors, had become an adversary (something like a landlord seeking to collect rent) who must be stalled and deceived and avoided at all costs. When eventually the true state of affairs became obvious, we were able to put her back on track towards her PhD, or rather, towards a much narrower PhD than we had originally anticipated. Her lack of engagement with us, when her problems first arose, meant that she came dangerously close to failure and certainly did not achieve her potential.

I don't understand this habit of hiding problems from a supervisor—it cuts off the possibility of getting help. I suppose it is a kind of fear of failure, a typical ego-protecting behaviour, but it reflects a confusion between the supervisor as a mentor and supervisor as an authority figure. Principally supervisors are the former, and for an experienced mentor some failure is hardly a surprise; we all fail sometimes. Concealing it helps no-one.

The cases of Hasrim and Delia concern problems that lay substantially with the student. Other cases of relationship failure lie with the supervisor. For example, occasionally a supervisor will lose interest in the student, or the topic, and no longer engages sufficiently well to ensure that the PhD stays on track. Such people have failed to recognize that the role of supervision entails more than simply getting a budding researcher to work on their pet project, but should in fact be focused on training and development, and that the actual topic may not be that important. Of course, all supervisors hope to advance their own work through the activities of their students, but this should be a secondary goal. Students who find themselves in such positions should explore all avenues, including fresh supervision, to ensure that their PhD goes well.

In short, having an open, positive relationship with your supervisor is a key part of success in completing a PhD; as is an understanding of your respective roles. This is one more aspect of research in which you need to seek a balance that suits your needs, but, once established, will help you to produce a strong thesis.

Summary of Chapter 4: Making a Strong Start

Starting your thesis:

- Write early, and write often. Keep your research in parallel with your writing so they grow in parallel. Begin to develop your thesis as part of the process of initiating your research. Create a table of contents as early as possible.
- If you do delay writing until after you have done your own work—although this is not the safest way to produce a strong thesis!—make sure that you are writing to the structure advocated above.
- Start with confidence. Write your introductory chapter first, then put it aside while you work on other parts of your study. Come back from time to time to revise your aim and scope so that they align with the changes you make as you go along.
- Let your writing drive your development of a literature review. Make sure that it is structured and critical. Use a rich mix of strategies for exploring the literature, including online academic tools, traditional libraries, and non-academic resources such as Wikipedia.
- In the early stages, your research questions may develop or change. This is a good thing.
- Some chapters are harder to write than others. A concrete chapter on your analysis, say, may be easy to produce and give you a sense of accomplishment; completing the background chapter will mean that the most difficult part of the thesis writing is behind you.

Within individual chapters:

- Start with an introduction that tells the reader why this chapter is included in the thesis, what you intend to achieve in it, and how you intend to do this.

- Develop the chapter in an appropriate and logical way to achieve the aim stated in the introduction. Avoid applying the same rigid template to every writing problem.
- Write a formal conclusions or summary section. Make sure that conclusions include a statement of the implications of the findings.
- Check that you have argued for the conclusions or findings in the body of the chapter.
- Check that these conclusions respond to the aim stated in the introduction to the chapter.

Writing strategies:

- Remember that it is your writing that will be examined. Other tasks may not be productive unless they lead to material for your thesis. Be aware of the tension between creating and critical thinking, and consciously exploit it to help you develop a strong thesis.
- There are many excuses for not writing. Most are a form of procrastination.
- Make plans, and stick to them. Audit yourself and seek to understand and resolve reasons for lack of progress. When you find an effective writing habit, use it.
- Make good use of your supervisor; think of her or him as a resource as well as a mentor.

Chapter 5
The Introductory Chapter

Imagine, for a moment, that your thesis is an important person you are meeting for the first time. It would be normal to be introduced as questions raced through your head: Where are you from? How did you get here? What are you doing here? What type of person are you? What have you done so far, and where are you going? As in social settings, I've noticed that students are sometimes in a rush to 'get started' and fumble the all-important first impression. Take some time to write the introduction properly, and revise it on a regular basis as your research project matures. Introductions are crucial, and it speaks for you as you strive to join an international community of scholars.

The purpose of the first chapter of your thesis is to introduce your work, period. I recommend that it be relatively short (seven to ten pages), and that it consist of five brief components: Context of the Study, Statement of the Problem, Aim and Scope, Significance of the Study and, finally, an Overview.

Establishing a Context

Your initial task is to situate your study so that readers can understand your specific points of concern. Before you begin talking about the problem, you must first provide some context to it. If you ignore this initial section, your readers may feel as if they have entered a conversation that has already started without them.

Of the many possible sources of inspiration that stimulate research, responding to a specific issue within an established agenda for research is perhaps the most common. By situating your work in this way, you immediately put yourself in dialogue with those in your field. Here's how Aek started his doctoral thesis:

From 'Chapter One: Introduction' in Phakiti, Aek. (2003). *An Empirical Investigation Into the Relationships of State-trait Strategy Use to L2 Reading Comprehension Test Performance: A Structural Equation Modelling Approach,* p. 1. Unpublished PhD thesis, University of Melbourne.

Broadly speaking, the language testing (LT) research framework stretches along a continuum. At one end, investigators assess and describe the language ability of an individual;

D. Evans et al., *How to Write a Better Thesis,* DOI 10.1007/978-3-319-04286-2_5,
© Springer International Publishing Switzerland 2014

at the other, they set out to establish standards for ethical conduct. Neither type of research, of course, proceeds independently of the other. Across this broad continuum of LT research, Bachman (2000) writes, five central strands have emerged: research methodology, practical advances, factors affecting language test performance, authentic or performance assessment and concerns with the ethics of language testing and professionalization of the field. Of these five, the present thesis is located within endeavors to examine factors affecting L2 test performance.

The style of writing that Aek has used reflects the style in his discipline area. By starting this way, Aek showed that he could identify key issues, understand the many strands of concern, and be focused on one particular challenge. From the start, the style and content of his work signalled to examiners and colleagues that he was set to productively engage in academic discourse.

Colleen took a different approach, starting her thesis with a deeply personal story:

From 'Chapter One: Thesis Introduction and Overview' in Nordstrom, Colleen (2004). *Beyond Modernist Instruments: A Reconceptualisation Of 'Quality Of Life' In Terminal Cancer*, p. 1. Unpublished PhD thesis, University of Melbourne.

... Months after arriving in Australia to start my PhD, my 38 year old sister, Rhea was diagnosed with terminal cancer. She was given only months to live. Without hesitation, I returned to Canada for several months to be with my older sister until she died. ... Returning to Australia, I knew that my research must be about terminal cancer, and it must be from the perspective of the person who is dying, not the health care professional. Not only did those months in autumn change my thoughts on death, it changed me as a person. Intellectually and personally, this thesis is my testimony of the illumination that has gradually transformed my life since embarking on the journey of trying to understand Rhea's experiences, and what it means to me, and how I approach my professional life.

Although Colleen could have written a formal introduction, she made a conscious decision to establish a more personal voice. She chose, consciously, to write in a style that, like Aek, signalled to examiners and colleagues that she was aware of the conventions in their community. Importantly, her style immediately aligned her work with the tenets of a qualitative approach that highlights the value of the 'lived experience' as a basis for empirical research. In a way, one aim of her thesis was to take such a stance as a way of questioning some of the assumptions that had dominated methodology in her field.

From the start, be aware of how your style of writing, your tone, and your use of key terms underpins your approach to research and provides you with an introduction to the research community. Should you write in the third person, like Aek, or in a way that is similar to Colleen? The best way to determine an appropriate style is simple: go to several top journals in your field and note the way that their contributors write. Imitate them.

Stating the Problem, Motivating the Study

Now that you've situated your work, your next task is to motivate further interest in the area—typically by 'stating the problem', because your research is nearly always an attempt to find a solution to a problem that you have identified. You believe that the present way of doing things is inadequate in some way, or that existing theory does not explain the observable behaviour of a system satisfactorily. Sometimes to call it a problem may be too strong: for example, if you were a historian you might be looking for a new way to view a series of events or the role of a particular person in them. In this case you might choose a milder description such as 'motivation for the study'. But don't omit this critical opening component. The examiner wants to know the driving force behind your research.

What should be in it? Certainly not a full review of the literature, although there might be some reference to it, because the unsatisfactory state of theory or practice might well be the problem—that is, shortcomings in previous approaches to the area might be the justification for carrying out the work to be described. There's no standard way to write a statement of the problem, but strive to make it relevant, definitive, and free of ambiguities. In many disciplines, especially those based in quantitative approaches to research, there is an expectation that the problem state-ment expresses the relationship between two or more measurable properties and thus can be empirically tested. The problem can then be written in the form of a hypothesis, or be stated as a single question such as 'To what extent do HEPA filters in ventilation systems reduce food contamination?' This could be posed as a de-clarative sentence followed by a series of sub-problems or questions, such as 'HEPA filters in ventilation systems can inhibit person-to-person transmission of airborne infections. To what extent can they inhibit person-to-food transmission?'

Over the course of your research, return to your problem statement on a regular basis and continue to refine it. With an eye on what you are discovering and the cur-rent research, ask yourself if your problem is still relevant, sufficiently narrow, and focused more on its underlying reasons than on approaches and techniques. How does it foreshadow and signal an overall approach to the investigation?

This is how Aek stated the problem that he had identified:

From 'Chapter One: Introduction' in Phakiti, Aek (2003). *An Empirical Investigation Into the Relationships of State-trait Strategy Use to L2 Reading Comprehension Test Perfor-mance: A Structural Equation Modelling Approach,* p. 7. Unpublished PhD thesis, Univer-sity of Melbourne.

To date, there has not been sufficient research that examines the connection between actual strategy use and actual L2 language performance. If strategy use does indeed make a dif-ference in L2 performance, it is equally important to understand the correlation between strategy use and L2 use performance. In summary, if we could systematically add all the pieces of information derived from research onto characteristics of the testing procedures, test-takers background characteristics and strategy use, it would be quite intriguing to know the degree to which these factors accounted for language test performance and the extent to which language ability *per se* was actually tested.

Note that there is no review of literature or theory here (he did review it later in his thesis), merely a clear statement that here was a large problem that was worth put-ting some effort into.

Grounded in a context, the statement of the problem sets out the reason the re-search is worth tackling. It acts as the precursor to the research aim. You will have to elaborate on the problem later in the document, for example, in the review of current theory, but not here. In summary, a 'Statement of the Problem' or 'Motivation for the Study' generally contains four parts:

- A brief history of the issue at hand ('Since the early 1950s, there has been …').
- A recent increase of the issue ('Recently, however, an increase in the …').
- Dissatisfaction with current knowledge ('To date, however, the lack of …').
- An identification of specific set of factors ('In particular, a focus on …').

Aim and Scope of the Study

Alastair was examining discrimination against the *burakumin*, an underprivileged group in Japan whose ancestors were outcasts because they worked in the unclean leather and butchery industries. He couldn't work out what he was trying to do in the whole project, even though he had written the background chapters and had prepared a paper for an international journal based on one of them. The following is the research aim he had written.

> The aim of the research is to establish which groups of mainstream Japanese continue to harbour anti-*burakumin* attitudes, analyze what those attitudes are and why they have remained extant, and to investigate which political measures are needed to solve the problem.

However, he had actually announced four aims in the same sentence:

- To establish which groups of mainstream Japanese continue to harbour anti-burakumin attitudes.
- To analyze what these attitudes are.
- To determine why they have remained extant.
- To investigate what political measures are needed to solve the problem.

What was the real aim? Almost certainly it was the last one. The other three were steps in the research method. He knew that he would not be able to make any suggestions about how 'the problem' could be solved unless he knew where it lay, and why it had persisted from the 1870s to this day. His problem statement would need to outline the problem step by step: that the Japanese government had formally declared these people to be outcasts in the 1870s; that their outcast status had become entrenched by custom until after World War II and that, as a result, they were discriminated against in education, employment, social welfare, and marriage; that the government passed legislation in 1969 to attempt to bring the social status of these people up to mainstream norms; but that, despite this, discrimination, although not as marked and no longer government policy, was still quite persistent. Thus the 'problem' was to find what an enlightened government might do next. Alastair needed to state that his aim was to solve this problem, not to tell the reader *how* he was going to solve it. The first three of his 'aims' should not appear until his research design chapter (see Chap. 7). And, having identified the problem, he needed to say what he meant by 'solving' it: in this case 'to identify ways in which the Japanese government could improve the social status of the *burakumin*'.

Your stated aim should have three characteristics:

- It should follow as a logical consequence of the problem statement. You identify a problem, and your aim is to address it; as just noted, you have to be clear about what the problem is.
- It should be singular. You must identify only one aim. This is not easy to do. Students often show magnificent ingenuity in stringing all the aims they want to include into the same sentence, as Alastair had. But four aims in the one sentence are still four aims. Nearly always some of the excess aims are in fact steps in the method that you have already unconsciously been working on to achieve that one true aim. If permitted to give just one piece of advice to students writing theses, I would say this: *stick to a single paramount aim.* If you do this, and get it right, the entire thesis can be built on this sole cornerstone. If you have two aims, you have to achieve both. To do this, you must address one first and then start another. Your thesis will be split in two parts, and these must be meshed into each other.
- The conclusions in your last chapter must respond to this aim. Obvious? Each of the examiners of my own thesis pointed out that I had promised one thing, and delivered another. Over the months you work on your thesis, it is easy to forget the relationship of the introduction to the final conclusion. When you have written the last sentence of your conclusions, go back and re-read your aim. If the conclusions don't respond to the aim, you had better rewrite it; and don't forget, you will also need to rewrite the problem statement that leads up to it.

Recall the list of questions that examiners may typically be asked to think of as they look over your thesis. One was, 'Does the candidate show sufficient familiarity with, and understanding and critical appraisal of, the relevant literature?' You will need to think critically at each stage of your work. To start the process of being critical, you must first set limits. A major part of being critical is to be able to set the terms of your debate and focus on what is particularly relevant to your aim.

To get students thinking, I ask an impossible question: What is the best restaurant near the university? Any answer sets off the need to ask a number of related questions. Best for whom? Under what conditions? How far away is it? What type of food? How expensive are the main dishes? You can quickly see that such an investigation could potentially never be completed. We then write out a scope for our investigation (for example, 'Find an Italian restaurant within five kilometres of campus that has meals for under $ 20') and start our discussion again with more focus.

In some areas of study, the scope of the investigation might require only a few sentences. In others, especially newly developing areas, it might require an elaborate discussion. (In an extreme case, the part of a thesis that has the most impact could be the discovery of a new problem.) Here is an example of a scope.

From 'Chapter One: Introduction' in Yeldham, Michael (2009). *Approaches to Second Language Listening Theory: Investigating the 'Top-down/Bottom-up Debate'*, p. 9. Unpublished PhD thesis, University of Melbourne.

Limits to the research are noted from the start. I do not, for example, examine the learners' listening development in interactional listening environments. The research is narrowed down to an emphasis on one-way, or transactional listening. The research is also focused on listening as an audio-based skill where no visual elements are presented. The study additionally focuses on 'learning to listen' rather than 'listening to learn' (Rost 2002; Vandergrift 2004). In other words, the emphasis is on learning how to improve one's listening ability rather than using listening as a vehicle to acquire the language. There is a need for these limitations because the investigation of listening comprehension processes and instruction is such a complex area.

In the study, I also limit the participants to native Mandarin Chinese-speaking EFL learners in Taiwan. I choose learners in an EFL rather than an English as a second language (ESL) context, because this EFL environment is more likely to minimise influences on the listeners' development outside the classroom environment, therefore providing a clearer insight into probable developmental effects associated with the instruction. Native Chinese-speaking learners are chosen because of the considerable distance between the English and Chinese languages, as English is a stressed-timed language and Chinese a tone language. This distance between the languages is especially pronounced in terms of the phonological differences between the two languages, which have been found to present major challenges for Chinese learners of English (Brown and Hilferty 1986; Pennington and Ellis 2000).

If you set out your limits, you are more likely to finish the thesis, your supervisor will know what you are interested in and resist attempts to send you further afield, and, most importantly, your examiners will be impressed that they have in hand a focused, high-impact study.

If you are working in an area that has a range of specific, but contested, words and phrases, you may need to set out a list of key definitions. In his thesis, Michael defined terms that are contentious in his field that include 'top-down and bottom-up', 'skills and strategies', and 'inferences, elaborations and guesses'. Because there are only a few terms, he explained these one or two pages into the introductory chapter. If he had had more such terms, and had to use a series of discipline-specific acronyms, I would have suggested that he put them in an appendix.

Significance of the Study

The scope is linked to the significance. A way to think of the significance of your thesis is to equate it with potential impact: Where do you think your study will make the most difference to current thinking? There are four primary lines of argument that may be used to establish the significance of a study. First, it may advance knowledge in the applicable field; that is, it revises or creates new knowledge (for example, the results will extend what is known about the applicability of a theory, the results are widely generalizable, or, for qualitative studies, transferable to other contexts). Second, a study may contribute to the solution of a practical problem faced by many others in the field (for example, the control of bacteria in food preparation, or the development of sustainable policies of food consumption). Third, it may demonstrate a novel use of a procedure or technique (for example, a statistical test, a projective technique, or an instructional procedure). Fourth, a study may contribute to part of a programmatic research effort; that is, when the results of the study are considered in relation to other studies, there may be theoretical or practical applications of major proportions. Each line of argument that is applicable should be pursued. Let's take a look again at Michael's work to see how he wrote about significance.

From 'Chapter One: Introduction' in Yeldham, Michael (2009). *Approaches to Second Language Listening Theory: Investigating the 'Top-down/Bottom-up Debate'*, p. 8. Unpublished PhD thesis, University of Melbourne.

One intended outcome of the study, on a theoretical level, is to identify a preliminary set of learner factors that arise out of extended exposure to two different approaches on the development of second language listening behaviours. On a practical level, a second intended outcome of the study is to clarify research techniques to do with the investigation of listening abilities. Specifically, I focus on Chinese learners of English at lower proficiency levels. Further, a third intended outcome contributes to the design of listening instruction by articulating a set of multi-level course design guidelines tailored for individual differences, particularly those associated with learner listening style. Evidence from reading and listening research supports this need to include both of these dimensions of listening proficiency and listening style in such a set of guidelines; findings from various studies (Davis and Bistodeau 1993; Vandergrift 1998a, b) have suggested the significant impact that both factors may have on learner response to differing forms of instruction relevant to the 'top-down/bottom-up debate'. Finally, a further concern in the research is the sequencing of skills and strategies for learners through a listening course.

Overview of the Study

The overview of the study should follow on logically from your statement of the aim. In other words, it is an annotated version of the table of contents. However, rather than writing it as a list of chapter and section headings, you should write it in the form of interconnected sentences and paragraphs to ensure that the logic flow is clear to the reader. That is, it should be a synopsis of the storyline that the thesis follows. Here is an example of an overview of a study.

From 'Chapter One: Introduction' in Al-Asmari, Abdul (2008). *Integration of Foreign Culture Into Pre-service EFL Teacher Education: A Case Study of Saudi Arabia*, p. 9. Unpublished PhD thesis, University of Melbourne.

This thesis consists of seven further chapters within three main parts. In PART I (Chap. 2 & 3), I situate the current study in related literature and establish the research methodology. In Chap. 2, I discuss the theoretical underpinnings of language and culture to guide the enquiry. This includes a critical review of the historical context, current practice, and the place of FC in practice and policy as well as issues relevant to integrating FC into EFL education. In this regard, I undertook a cross-disciplinary approach drawing on areas such as language policy and Higher Education, World Englishes, Intercultural Language Learning (ILL), and Computer-Assisted Language Learning (CALL). Combining these insights, Chap. 2 argues for the need to investigate the place of FC in pre-service EFL teacher education. Based on that, the most pressing gaps in the literature are identified and research questions are posed accordingly. Chap. 3 deals with the methodological issues and research design providing the philosophical foundations, case study context (i.e. SA), theoretical and procedural description of instruments used in the study to collect, present, and analyse data.

In PART II (Chap. 4, 5, & 6), I present the results of data analysis of the place of FC in policy, instructors' practice, and pre-service teachers' EFL learning respectively. PART III (Chap. 7 & 8) contains the discussion, recommendations, and conclusions of the current study. In Chap. 7, the discussion on the key findings is expanded and appropriate concepts to integrate FC into EFL pre-service teacher education are developed. Based on that, implications and recommendations are drawn to inform the policy and practice of pre-service SEFL teacher education. Finally, Chap. 8 contains the conclusions and reflective evaluation of the study and suggests further research agendas.

An examiner who read this would have a clear idea of not only how the thesis was going to develop, but also of how the various parts of it relate to each other.

I suggest you call this section 'Overview of the Study', rather than 'Method' or 'Research Approach', even though it describes the method that you will use for your whole research project. The difficulty is that most people use the word *method* to describe the method used in the part of the total research program they have designed themselves—their own surveys, interviews, observations, analyses, experiments, and so on. They then often give Chaps. 4 or 5, which introduces this research program within a research program, the heading 'Method'. The point is that you are in danger of using the word in two senses: the method used to develop the whole of the report or thesis, and the method used for that part of the research program that you designed yourself. I recommend that you reserve the word 'method' specifically for the part of the work designed by you, the researcher, and to use *overview of the study* to describe the approach used in the whole project—which will include historical reviews, reviews of theory and practice, accounts of the researcher's own work, and synthesis of all of these to permit conclusions to be drawn.

Research Questions and Hypotheses

Should research questions and hypotheses go in the first chapter? Some supervisors would say that it was an absolute necessity; others that they should only be stated after the background, and not before; others would tell you that they are not required at any stage of a thesis (but they're wrong). As you read other theses, and become a member of a specific research community, you can make up your own mind on this matter. If you asked me, however, I would tell you to at least make a general statement of the research questions or a hypothesis in the first chapter, or alternatively, make sure these are clear in the statement of your aim. Otherwise the examiner may struggle to understand what the point of the thesis is, and an unmotivated reader is an unhappy reader.

It is certainly true that research questions can only be generated after an area is reviewed and gaps are identified. For many of my students, this requires that they read widely and, in the process of reading and writing, sharpen their critical thinking skills. They may even need to conduct a preliminary investigation to further tease out and clarify questions. Accordingly, the right place for a precise statement of the research questions may indeed be at the point where the background chapters finish, and just before the start of the core of your own work, but the questions need to be stated in broad terms in your first chapter.

Now, as for hypotheses, let me tell you about Daud, who came to me wishing to do a Masters. When I asked him what the aim of his project was, he replied that it was to demonstrate that development in his home country, one of the African nations, depended crucially on the adequate provision of household energy supplies. This is a hypothesis, not an aim. (An aim here might be to investigate the relationship between national development and the availability of household energy supplies.) This confusion is very common. More often than not, when I ask potential research students what the aim of their research is, they reply with a hypothesis. The confusion seems to be due to a looseness of expression among research workers when talking about research. As discussed in Chap. 2, research is a complex mixture of creative and rational processes. As a result, it is quite common to leap right into the middle of the research process with a hypothesis, and work backwards to the aim and forwards to the conclusions at the same time.

However, no matter how irrational and chancy an actual investigation is, the output of research must be written such that it is argued logically and clearly. Therefore you must eliminate any confusion between *aim* and *hypothesis*. An *aim* is to do with directing something towards an object, whereas a *hypothesis* is a proposition made as a starting point for further investigation from known facts.[1] Clearly the two words have quite different meanings, and should not be used interchangeably.

To be fair to Daud, he was not really confusing the two things. He *was* giving me hypotheses, or propositions that could be tested. He had perceived problems, and had developed hypotheses about them in his unconscious thoughts over a period of time, long before he had come to me to propose a plan for research. When he came to see me he had not yet worked his way back from a hypothesis to an aim. He was focused on what he could *do* in his study, rather than what he was trying to *achieve*.

When I explain the difference, referring to the dictionary if necessary, students often reply that their aim is to 'prove' their hypothesis. This is not an aim either! Proving is what we do to hypotheses, at least in the sense of 'proving' as testing (as in 'proving ground'). A hypothesis is a device that enables researchers to set up useful tests or experiments that will tell them whether they are on the right track. It is not the arrow pointing to the destination.

A suggestion, then, is that you not use the word 'hypothesis' in the opening chapter of your thesis. In a recent seminar a student told me that her supervisor said

---

[1] GW Turner (ed.), The Australian Concise Oxford Dictionary, Oxford University Press, Melbourne, 1987. 'Consider an archer: her aim is to put the arrow in the bull's eye. She might have a hypothesis that if she was shooting in a northerly direction and the wind was blowing from the east at 10 m per second she would have to shoot at a point 2° to the right of the bull's eye. She could quite easily test this hypothesis by shooting a group of arrows at the bull's eye and another group 2° to the right. When she had tested her hypothesis she would be in a better position to achieve her original aim, which was to get arrows in the bull's eye.'

the first chapter of a thesis should be an extensive review of the literature ending up with a research hypothesis, which seems to cut across my advice. However, as pointed out in Chap. 2, what he was really advocating, without realizing it, was omitting the first chapter of the thesis altogether. This is dangerous! If you do this you will inevitably get the aim, research questions and the hypothesis mixed up. As a result, you run the risk of never identifying the purpose of your thesis clearly.

Your first mention of the word hypothesis should be in the chapter concerned with your own work. When you do use it, stick to its formal meaning. You should be able to deduce from the combination of your review current theory and practice, and possibly from your own preliminary studies, that there are certain lines of thought that are worth investigating by careful tests. These tests are the ones that, in the classical 'scientific method', are called crucial experiments. They will tell you without doubt whether your hypothesis has stood up or whether it has been demolished. Either way you have made progress.

Revising the Aim, Aligning the Conclusion

It is necessary to revise this introductory chapter as you make new discoveries in your work. Almost imperceptibly, your aim can change as you go along, or your work may no longer lie within the scope you first established. When I wrote my thesis, the first draft of my introductory chapter was essentially a summary of my first go at understanding the literature. It turned out to be over twenty pages. Later, I trimmed it to a much more focused and manageable seven-page chapter. One of the holdovers, however, from my earlier attempts of writing out the introduction was that I had made a 'promise' that a key implication for my work rested in classroom practices. During the course of writing my thesis, my focus became much more theoretical than I had originally planned. Unfortunately, I failed to delete my earlier concerns from my introductory chapter. The examiners noted this point and wondered why I had not addressed it in my final discussion. When I re-wrote the first chapter after examination, the first thing I did was to delete any suggestion that was work was focused on something that I had failed to discuss beyond the first few pages!

Many students resist writing the introduction early in the project because they suspect that its direction is likely to shift as they develop their thinking. Don't resist, or delay and let this stop you. Write the introductory chapter as confidently as you can when you first start out. In this first draft you may not see the need to put any limit on the scope of your study, and almost certainly you will be a bit tentative in your aim. But you will have made a start. As you come to grips with the complexity of your research, revisit the introductory chapter periodically and make minor changes; or just read it over, and make sure that it still applies. Are you still trying to achieve a particular *single* aim? Are you still within the parameters of the scope that you first set out? One way to see if you are drifting from your original thoughts is to look over your working title. Have you thought about changing the title, and does it still describe what you are doing? I suggest that you have a good look at your introductory chapter at regular intervals, say every few months, and certainly whenever you complete another chapter.

In thesis writing seminars, I've also been giving another bit of advice: on the day you start your introductory chapter, start a conclusion chapter too. I advise students to work on the introduction and the conclusion throughout their research. Use the two chapters to provide banks to the flow of your growing argument. Students are often surprised when I offer this advice. How can I begin to finish when I'm just starting? I then remind them of the old advice on travelling: If you don't know where you are going, you will end up somewhere else. Exploration and meandering are fine for the casual tourist seeking to explore a new country, but not for a determined research student. Start with the end in mind, know where you are going, and work backwards with an eye on the clock, a list of resources, and a compass in hand.

Be aware that you will vary your purpose, change direction, and reconsider the scope of your study (it will happen!). Each time this occurs, use revisions to your introductory chapter to document and acknowledge those changes. At the same time, go to the conclusion chapter and revise the aim. Keep the beginning and the end in alignment. By doing this, you won't wake up one day to discover your thesis has 'gone off track' and no longer meets its stated purpose. Best of all, however, you will prevent the possible distortion of your own structure. Remember, although you wrote the thesis over as long as a period of years, your examiners read it over a period of days, and any structural problems will be very obvious to them.

Summary of Chapter 5: The Introductory Chapter

Your introductory chapter should consist of five brief elements:

1. Context of the Study

   - Provide a brief history of the issues to date.
   - Situate your particular topic within the broad area of research.
   - Note that the field is changing, and more research is required on your topic.

2. Problem Statement (or Motivation for the Study)

   - Identify a key point of concern (for example, increasing use or prominence, lack of research to date, response to an agenda, a new discovery, or perhaps one not yet applied to this context).
   - Refer to the literature only to the extent needed to demonstrate why your project is worth doing. Reserve your full review of existing theory or practice for later chapters.
   - Be sure that the motivation, or problem, suggests a need for further investigation.

3. Aim and Scope
   - Be sure that your aim responds logically to the problem statement.
   - Stick rigorously to a single aim. Do not include elements in it that describe how you intend to achieve this aim; reserve these for a later chapter.
   - When you have written the conclusions to your whole study, check that they respond to this aim. If they don't, change the aim or rethink your conclusions.
   - If you change the aim, revise the motivation for studying it.
   - Be sure to establish the scope of your study by identifying limitations of factors such as time, location, resources, or the established boundaries of particular fields or theories.

4. Significance of the Study

   - Explain how your thesis contributes to the field.
   - There are four main areas of contribution: theory development, tangible solution, innovative methods, and policy extension. One of these contributions must be identified as the basis of your primary contribution to the field.
   - In contrast to reports for industry, theory development is an expected and required contribution; for PhDs in particular, it must be 'original'.

5. Overview of the Study (or Structure of the Thesis)

   - Sketch out how the thesis is structured. Don't confine yourself to a list of the chapters, but show how they are linked and that one section logically leads to another.
   - Check whether the reader can see from this sketch how the aim will be achieved.

Revising your introduction, aligning your conclusion:

- To avoid making promises to the examiner that you can't keep or that you do not later address, regularly review your introductory chapter and revise it accordingly.
- Consider starting your conclusion chapter on the same day that you begin your introduction, and, each time you work on one, work on the other, thus keeping them in alignment.

Chapter 6
Background Chapters

Depending on the nature of your thesis, the background sections or chapters can take any of several different forms. However, their functions are always the same: to provide the context for your own work and to be the starting point for an examiner to think about your position in relation to the work, that is, to be the 'you' that you were at the start of the research project. The four most common elements in the background chapters are:

- *Establishment of a context* to locate a study in time, location, or culture.
- *Identification of current theory, discoveries, and debates*, including an evaluation of those most useful and salient to your topic, as well as a nomination of gaps in literature.
- *Understanding of current practices and technologies* in your field that highlight, and perhaps synthesize, a selection of the appropriate methods to gather data for your study.
- *Preliminary investigations* done by you or others to help clarify research techniques, formulate hypotheses, or focus areas of investigation for the major research program to follow.

You may have one or more of these, depending on the type of research you are doing. For example, if you were investigating ways that vitamins have been promoted as a regular dietary supplement, you would certainly need a descriptive section—or, perhaps a full chapter—on the development of the understanding of vitamins. You would also have a section (or chapter) that sets out many of the current controversies that surround the taking of food supplements and surveys a range of expert studies on the topic. Next, you would need to have a section (again, or chapter) that details ways in which studies concerning the promotion of vitamin supplements have been conducted. If these are unsatisfactory in some way, for example, questionnaires in them are out of date or not applicable to your intended participant group, then you would need to conduct a preliminary study to clarify instruments and other aspects of methodology, including research questions. These background sections, or chapters, would provide the basis for the design of your own work, which then could be understood in context, located within a theoretical framework, and conducted using techniques backed with solid justification.

D. Evans et al., *How to Write a Better Thesis*, DOI 10.1007/978-3-319-04286-2_6,
© Springer International Publishing Switzerland 2014

Developing Critical Thinking

One approach to thesis writing requires students to review the literature and produce a chapter entitled Literature Review', thorough and polished, before they are permitted to proceed with their own work. The idea is that, informed by a literature review, they would be able to see just where previous researchers had drawn unwarranted conclusions or had disagreed with each other, and would then be able to design brilliant experiments to resolve these problems.

You certainly should read the literature before you leap into a full-scale research program, and should also attempt to write down your understanding of it. This is a good way to learn how to follow the important arguments through, and to understand the agreements and disagreements. But until you have done some work of your own—perhaps collected and analyzed some data, for example—it is not possible to be 'critical' in the sense implied by 'a critical review of the literature'. It follows that you will not yet be able to design those brilliant experiments that have so far eluded the other researchers in your area. You may be able to design a research program, but almost certainly it will be tentative. With luck, the results of this preliminary program may help you to design a better set of surveys or experiments next time (by which time you will have thought about it a bit more, and will have gone back and re-read the literature). It also follows that your reviewing of the literature is an ongoing process. You should still be reading it when you are setting out results, discussing your findings, and writing conclusions.

How do you convert your initial 'literature survey' into a critical review of existing theory that will lead logically into the work that you design and undertake yourself? A review of current theory serves three purposes: it gives the background information required to contextualize the extent and significance of your research problem; it identifies and discusses attempts by others to solve similar problems; and it provides examples of methods they have employed in attempts to get these solutions. Make sure you deal with all of these.

The first purpose is the most straightforward. Its sole purpose is to establish the parameters of your argument. Guide yourself through this section of the research by asking the four standard journalist's questions: Who? What? Where? And when? Keep this section short, and do not get caught up in unnecessary detail. Simply put, it provides a map of the territory you are seeking to cover. It signals to your readers that you intend to follow the scope of your investigation and are confident enough to guide them through the complexities of the topic.

First attempts to review existing theory often stop after an initial draft. But when you have put your problem in the context of ongoing research in the area you have hardly started! Identifying and discussing possible solutions to your problem is the second purpose of your review. This is where you need those critical skills. It is likely (and expected) that you will have read much more widely in the topic area than you need for your review. Your initial journey through the literature will have helped you to gain a better understanding of the many complex facets of your central problem, but you do not need not to write about all of them in full. Keep in mind the aim and scope of your thesis: How does what you are reading relate to achieving these?

As you read, write: the act of writing forces you to come to grips with conflicting ideas and focus your attention on the most important arguments. Eventually, you will gain a sense of what parts of the previous research are leading you towards possible ways of dealing with your problem. As you develop a stronger sense of the field, strive to filter the good from the bad. What you are doing at this point is creating an internal set of criteria on which to accept or reject arguments and, through this process, you are developing the skills of critical thinking. By now you will probably have written many fragments and mini-reviews, and it is time to write a serious first draft of your 'critical review of existing theory'. Before you triumphantly hand this to your supervisor for criticism, it's a good idea to put it aside for a week or so and work on something else. Then come back to it and try to rework it into a second draft in which you attempt to articulate the criteria you have been developing, and demonstrate to readers just how sharp your criticisms are. Share your work with colleagues and read their work too.

That is, you should read, and think, like an examiner. With experience, researchers accumulate a toolbox of questions that they use to evaluate the work of others, and of observations of common ways in which other work is flawed. Consciously building this toolbox can help us become better at critical thinking.

Effective critical thinking depends on effective reading. For me, reading a piece of research literature seems to fall into phases. The first phase, counter-intuitively, is fairly uncritical. I try to get a sense of what the researchers were trying to do and whether the problem is genuinely interesting,[1] and then to understand how they undertook the work. Once I have a broad grasp of what a paper is about, I begin to look at issues such as whether the results really support the conclusions and whether the experiments look robust. A big question is whether the work is significant; some papers are genuinely remarkable, but most are an incremental contribution and need to be analyzed from that perspective. In considering whether the work is dependable, it also helps to consider the reputation of the authors, which may seem unfair—anyone can do great work—but a senior researcher is unlikely to knowingly put their name to flaky or insignificant work, while a more junior researcher may be desperate for any kind of publication.

Some papers are plain wrong or misguided. The fact that they are published means that someone believed in them, and it is certainly the case that high-impact journals are more trustworthy than fringe publications, but you should always be sceptical. It is up to the author to convince you that the work is correct. At the same time, a paper can have strong results even if you don't understand it.

---

[1] All too often the answers to these questions are 'they don't seem to know themselves' and 'no'. I'm not kidding—it is remarkable how often researchers don't appear to have a clear idea of why they are doing a particular piece of work.

Establishing Context

Tony was writing a thesis on the usefulness of laser grading of irrigated farmland. In this process, laser-guided machinery is used to grade the land to high precision, so that the irrigation water will flow evenly, the depth of flooding required to ensure that the whole area is properly irrigated is reduced, and no areas are left water-logged. These good effects may be counterbalanced by alterations to plant growth caused by disturbance to the soil by the grading operation, for example, through loss of soil nutrients and bacteria. To understand all of these effects, the reader needs a description of the interaction between soil, water, plants, and air.

How much of this should Tony have included as a background chapter? At first he left it all out, assuming the reader would know as much about this as he did. I, his supervisor, didn't, and told him that he would have to include enough of this material so that the policy-maker, as distinct from the agricultural scientist, would be able to understand what he was talking about. A week or so later I asked how his description of the soil and plants was going, and he told me that he had written about 30 pages and was only halfway through. He was busily paraphrasing from half a dozen standard soil texts and distilling the thoughts of the three soil scientists he had previously interviewed. What he should have been doing was something between these two extremes.

How can you determine what to omit, and what to include, as you establish context? I suggest three rules:

1. Don't include material that the reader does not need in order to understand what will follow. Although we need some chemistry to understand the effects of bicarbonate soda in baking, there is a lot that is not relevant to the problem.
2. Don't include anything in your main text if it is going to interrupt the development of the flow of logic in your argument. There may be some things that have to be included in the thesis, but that should be in appendices rather than the main text.
3. Include anything that is genuinely clarifying.

The 95 % Syndrome

As students get further and deeper into their projects, they often fail to realize how expert they have become in their area. They have absorbed the key ideas that dominate their particular field and have come to take them for granted. When they start writing about their own research, which is about the extension and modification of these ideas, they assume that the reader will be just as familiar with the basic ideas as they are, and they don't bother to explain them properly. They assume the 95 per cent and concentrate on the 5 per cent.

What a mistake! First, few people, anywhere in the world, will be as familiar as you are with the basic ideas. Who else has spent years studying this same combination of topics and questions? Second, although their readers, the examiners, may be assumed to be generally familiar with the field of the work, they will not necessarily be expert. If there are two examiners, often one will be chosen because of familiarity with one aspect of the work, and the other examiner because of another aspect. If there is an examination board, some may be chosen because of their experience rather than their knowledge, and all examiners are chosen more for their academic credentials and ability to judge critical thought than for their detailed knowledge. Critical thought will include, among other things, a clear, open-minded attitude to existing thought on the topic, rather than a tendency to rigid judgments.

So, don't ever assume that the examiners will know most of what you have learnt during your project, and that you have to discuss only the remainder that you believe is innovative and challenging. As I said at the start of this chapter, think of the old 'you', and what you knew before you got started on this particular research topic. That is the person your background is to be written for.

A related issue is that you need to learn to write *defensively*. That is, if you think something might trouble an examiner, then address it. If they might confuse your topic with some other similar but fundamentally unrelated topic, help them by explaining the distinction.

Defensive writing, though, requires a certain amount of mental gymnastics. It means that you have to ask questions like, 'If I were a reader, and I read this carelessly, or I didn't really know the background literature, then what mistakes would I make?' And then you need to change your thesis to help the reader avoid these mistakes. But, as in so many things in research, the exercise of forcing yourself to take a different perspective—of trying to see your work from someone else's point of view—can be highly productive. It is particularly valuable when you are trying to shape the context, but is useful for every part of your thesis.

Understanding Current Theory, Discoveries, and Debates

Examiners will be sensitive to instances in which major contributions are neglected, or their significance downplayed. Summarize their contributions completely and honestly. But remember also to point out how these other studies may have advanced the discipline. For example, one student of mine, Raymond, had a tendency to write about all previous papers as in one of two classes: a few papers were insightful, groundbreaking, and of critical importance; the rest were, in his view, more or less misguided, confused, foolish, or wrong. He often failed to see how they made useful contributions (perhaps in a context that was now outdated, which however does not mean that the work was invalid), possibly because of a lack of appreciation of the fact that much research is incremental. At times he almost seemed to want to be a giant-killer who was bringing down the inflated reputations of esteemed researchers. The net effect was that his criticisms could seem inconsiderate and harsh, that is, they lacked balance.

Throughout this second section of your review, keep in mind that you are engaging in a conversation with other academics. Engagement is the key concept: it is a spirit of 'give and take' that respects the value of multiple perspectives. It is easy to make the mistake of thinking that the function of this section is merely to 'report' or 'describe' previous studies in an effort to show that you have 'done your homework'. Rather, you should interweave various studies to build up the argument that the problem you are tackling is not yet solved and still raises some interesting and unanswered questions.

Eventually, you will come to an understanding of the most recent thinking in the field. At that point, briefly summarize the main points that are still troublesome. You have identified the 'gaps' in the theoretical framework and areas that have remain relatively unexplored by previous researchers. This summary should be setting the ground for the questions or hypotheses that you will be identifying in your chapter on design of your own research. In a sense, these are the gaps that you are trying to fill with your own original contribution.

Understanding Current Practices and Technologies

In a third section of background material, you will need to examine the approaches and techniques others used in research in your topic area. From your previous reading and your attendance at research seminars in your department you will probably have become aware of the flaws in research design, research methods, and the reporting of results that can mar an otherwise competent investigation.

Once you get to know some of the common mistakes, stay alert to them as you review previous studies. Use your knowledge of such mistakes to point out how a previous investigation may have made only a limited contribution to solving the problem you have posed. (If the contribution had not been limited, you would not have to conduct an investigation—somebody else has already solved the problem!) Where appropriate, point out the limitations to those approaches; remember, one of the attributes of a PhD thesis is that the student is aware of limitations. No method is 'perfect', but your review should lead to an understanding of which methods can be used to achieve solutions to your problem. You will need to draw on this in your next chapter, where you select appropriate methods for your own research program.

In some disciplines, a key purpose of this part of the literature review is to establish *baselines*. The aim of your work may be to improve on the state-of-the-art: a more sensitive test for contamination, for example, or reduced energy consumption in food preparation. The very presence of these comparative terms here (improve, reduce) suggests that there is something you are comparing against. This is your baseline. You need to know what the current best competitor is, and later you will need to test whether your approach has advantages. Perhaps obviously, there is little value in showing that your approach is better than something which is already known to be poor—so you need to use your literature review to show that your chosen baseline really is the best that is currently available.

You may then even undertake some preliminary investigation of this baseline, to establish its strengths and limitations, as I now discuss.

Preliminary Investigations

Peter came to me in some puzzlement, announcing that his experiments had failed. He was investigating a problem in the utilization of brown coal, and had designed some preliminary experiments based on hypotheses drawn from earlier work in the same field. What he expected to happen had not happened, and what had happened was quite unexpected. His 'experiments had failed'. I assured him that they had in fact succeeded. He had been lucky enough to have the unexpected happen, under conditions where he was *sure* it had happened, because he was making careful observations. If he could work out why it might have happened, he would have far better hypotheses than when he was depending solely on the earlier literature.

Most research workers when tackling a new project will use a variety of methods. As you can see from the example of Peter's work, it can be difficult to establish from the literature alone what experimental work you should do, because you are an outsider listening to a debate. If you are to become a participant in the community of practice then you will need to have some practical experience of your own. In the physical or biological sciences this might consist of designing some simple experiments to enable you to test the results or theories of earlier workers, as Peter had done. In the social sciences, gathering data through preliminary surveys or interviews could be useful. A further method is to revisit old questions or experiments in the light of new methods and materials.

Where should an account of this preliminary work appear in your thesis? If you have used it to help you to formulate hypotheses that you have called on when designing your principal research program, you could report the preliminary work as one of the background chapters. If it appears to form a major element of the principal work itself, you should set it aside for reporting later as part of the 'Design' and 'Results' chapters. If you report preliminary investigations in a background chapter, it will have to contain sections on the hypothesis used, the design of the work, the results, and the conclusions drawn from them. In either case, be sure that you make clear the need for its inclusion—you don't want to appear as if you are trying to pad out your work by including irrelevant material.

Revising

As I recommended earlier, there is much to be gained in writing the background chapters before or during the time when you are carrying out your own research program. However, when you have finished your own research it is time to rewrite the background chapters. You are now much clearer about several things than you were when you first wrote them:

- You understand the links between your own work and the work of others who went before you.
- You now know what assumptions you made, perhaps unconsciously, about your study area. These can now be made explicit.
- You are aware of the issues surrounding the application of current methods in your field, and have explicitly pointed out their limitations.
- You may have realized (perhaps with the help of feedback from your supervisors and participants in your research seminars) that you were making unwarranted assumptions about the level of knowledge of other people about the background to your own work. For example, if you come from India and your project is located in India, you might assume that the reader was as familiar with the names of the states of India as you are. What if one of your examiners comes from Finland?
- In your efforts to understand and interpret the results of your own work you will have reached a new level of understanding of the work of others—this is what is meant by a 'critical' understanding. For example, if you were working in the area of 'sustainable architecture', you would by now have realized that many people writing about it had been using the words as a vague catchphrase, and you need to go back and make some careful definitions.
- Most likely you designed your own work without being completely conscious of the research questions or even the hypotheses that informed it. This possibly sounds silly, but my experience with students is that this is what often happens. When I ask students why they did something and they have no ready answer, they sometimes seem shocked. This does not mean that the work does not have a valid basis, but rather that the driving force behind it was in part developed in the unconscious mind. It now has to be made explicit. In the chapters on your own work (see Chapter 7) you need to make it explicit, and your background chapters will have to lead into your research questions or hypotheses.

This gives us a framework for how you should tackle the revising of the background chapters, as explained below.

Ensure that the ways that you are going to use words and ideas are carefully defined. Where these are fundamental to your own work, the development of these ideas in the literature or even in the history of ideas must be discussed. For example, in her thesis on landscape heritage, Jan had to trace through the development of the notions of heritage and landscape. Both of these words have a host of everyday meanings, but it was fundamental to her research that the reader understood precisely how she was using the terms.

Any formal literature review that you do before you begin your own work should not appear in the final thesis in that form. What you should have, rather, is a structured account of the literature that is current at the time you did your own work. You will be able to impose a structure on it, because by now you have largely finished your own work, which will have gone further than the work of others (that's why you were doing the research). You can write it with the critical perception of the worker who has now gone past this point. The story in Chaps. 2 and 4 about Karen's work is a good example. She couldn't get started on her review of existing theory because she felt she had to describe the work appearing in 120 different papers. When she structured it around the central four ideas that represented significant advances in the field it was easy to write. You should also ensure that the structure provides a firm base for the discussion of your own work (see Chap. 9).

There needs to be an appropriate formulation of the research questions or hypotheses that you used to help you to design your own research program. When students present me with a proposed work program and I ask what it is based on, they sometimes reply that it is obvious, or that it just came to them. These responses may be true from where they stand, but will not convince examiners. They have to be argued out. What apparently happens is that our unconscious mind works on various fragments of ideas from different sources that come to us from our reading and our senses, and makes connections that our rational mind will not. These connections emerge not as new rational thoughts but rather as proposals for action. We then implement these proposals in the form of research designs without actually making the underlying logic of them explicit as research questions or hypotheses. If this happened for you, you now have to work backwards from your research program to why you did it the way you did, that is, what your research questions or hypotheses were. You then have to work back further to where these questions or hypotheses came from, and ensure that at the very least the conclusions of your background chapters prepared the way for them. You then have to make sure that the appropriate material is present in the background chapters to enable these conclusions to be drawn.

You have to be ready to cut material out of background chapters if it is not used elsewhere in the thesis. The background chapters are not an end in themselves, they are merely the context for your own work. I have already mentioned some tests for what to include and what not to include in descriptive chapters. Be ruthless! If you are not making use of material either as background to your own work or as context for the discussion of your results, take it out. A survey of literature on a topic unrelated to your own work will not please an examiner looking for evidence of critical thinking.

Finally, there is another type of material that you should remove. Because students know what they found out themselves they sometimes forget that the examiner does not know. Remove all material from background chapters that is to do with the design of your own work, the results of your own work, or the discussion of these results.

Summary of Chapter 6: Background Chapters

The background chapters have the following two functions.

To provide all the background material needed for your own research:

- Historical, geographical, and other descriptions of your study area.
- Definitions and usages of words and expressions as appropriate to your thesis.
- Existing theory and practice for your research topic.

- In some cases, preliminary reviews, surveys, correlations, and experiments following what other workers have done.

To provide the stepping-off point for your own research:

- The conclusions to these chapters should lead clearly to the research hypotheses or research questions that you pursue in your work (see Chap. 7).

Writing your background material:

- Write first drafts in the first year of your project. Use the style, referencing, and so on that you intend to use in the final thesis (see Chap. 3).
- Many researchers, particularly in the experimental sciences, put this writing off until they have finished their research. Don't delay. Writing early drafts helps to sharpen up your research design.
- These first drafts will probably not be well structured, as you are not yet on top of your topic. Be prepared to restructure them later, after you have done most of your own work. This double handling is not a waste of time, as it will make a fruitful contribution to your own research.
- As you revise, make sure that the background does lead into your research questions or hypotheses.

What you should include:

- All necessary definitions and ways that you use words or ideas in your own work. Don't assume that the examiner will know this. This is particularly important in cross-disciplinary research.
- All the necessary geography, context, and history.
- All the arguments that are in the literature, and some tentative judgments on where you stand (but don't enter the argument yet; wait until you have described your own work).
- Everything necessary to justify the conclusions or summaries to the chapters, which in turn have to lead to your research hypotheses or questions.

What you should not include:

- Descriptive material that will never be used later in the thesis. Your first draft may contain a lot of this. Be ruthless: take it out!
- Your own contribution to thinking on the theory. By the time you come to revise these chapters you should be in a position to make such contributions. Resist the temptation, and save these contributions for your discussion chapter.
- Any foreshadowing of what you will be doing in your own research. You can't do this until you have designed your own research, which can't be done until you have finished all these chapters. Don't get ahead of yourself.

Chapter 7
Establishing Your Contribution

Think back to when you began your research project. In all likelihood, you wanted to do research because you were intrigued by or eager about something. Perhaps your ideas were vague or ill-formed, and even possibly you were happy to join any existing project in a broad area. But soon you developed a definite problem that you were working on, with the intention of making a contribution to your field.

The part of the thesis I discuss in this chapter is where you begin to present and argue for that contribution (though I would be stunned to see a thesis chapter with that heading). In the introduction and background chapters, you established the setting on which the contribution rests; in the results, discussion, and conclusion chapters to follow, you will demonstrate the extent to which your contribution is correct, complete, and significant. But let me stress that I use 'contribution' here only as a convenient label to cover a wide variety of alternative kinds of structure and content; and also that, in some respects, the whole thesis is a contribution, from your novel analysis of the background literature to your approach to interpretation of results.

Also, observe once again that in some disciplines—particularly disciplines where the research is progressively published during the course of the project—a larger thesis such as a PhD may be made up of a series of linked contributions rather than consisting of a single consolidated piece of research that resolves a single question. In such disciplines, the 'contribution' may be spread across several chapters, each with a self-contained mini-thesis–like structure of research question, innovative proposal, results, and analysis. For convenience, the discussion here is of a thesis with a single contribution, but the guidance applies to any kind of thesis structure.

Another disclaimer: the diversity of contribution in different theses means that any specific advice on how to present this chapter or sequence of chapters is likely to be wrong or inappropriate. This is an area where you need to be led by examples from your own discipline. There are however good general principles to observe, as I now discuss.

D. Evans et al., *How to Write a Better Thesis*, DOI 10.1007/978-3-319-04286-2_7,
© Springer International Publishing Switzerland 2014

Kinds of Contribution

I continue to be amazed by the breadth of *kinds* of contribution that research students can make. That is, not only do they work on a remarkable spread of questions, but even the nature of the questions varies dramatically. To help explain what I mean, here is a sample of research contributions that I found in my university's PhD thesis collection:[1]

- Design of health messages for display on the packaging of fatty foods.
- Historical food advertising themes as a measure of population dietary changes.
- Food choices in literature as signifiers of masculinity.
- Historical marsupial dietary adaptations in response to climate change.
- Marsupial dietary adaptation mechanisms for coping with climate change.
- Practical tests for detection of salmonella in the home.
- Low-energy fabrication of food-storage plastics.
- Atomic models of carbohydrate-based metabolism.
- Emergency-care identification of dietary causes of rapid-onset illness.
- The effectiveness of taxation for reducing dietary causes of type II diabetes.
- Statistical minimization of food wastage in complex distribution networks.
- Energy accountancy practices for production and distribution of grains.
- Diet in Camanderra: contrasts between rural and metropolitan Australia.
- Quantification of the impact of diet in treatment of recovery from stroke.

These truly are diverse. Some are theoretical. Some are based on data collected over time, prior to the research being done. Some are based on a new investigation of historical materials. Some are based on data gathered for the specific purposes of the project. Some are analyses of other people's proposals or initiatives. Some propose a new technology, or treatment. The kind of evidence that can be used to validate the work varies too.

Very probably, none of these projects is close to your own area of expertise, but you may find it interesting to work through them and try and characterize them against a range of properties. Does the work involve inventing something new, or observing something that already exists? (Example: new plastics are an invention, quantification of food choices is an observation.) Is the impact limited to the research question, or is it likely to have broader implications? (Example: marsupial biology may not apply to other animals, but diet in Camanderra may be representative of regional Australia in general.) Is it primarily theoretical or practical? And so on.

The lesson here is that each kind of contribution is likely to be suited to a particular style of narrative. Your task is to figure out the narrative that best explains your contribution.

---

[1] This is a good place to remind you that my descriptions of research are to some extent fictionalized. That is, except where a citation is given, I have used real students and research questions, but have altered the research descriptions in ways that I feel preserves their 'feel' while making them both anonymous and more accessible to a general reader.

To arrive at a narrative, it helps to have an understanding of how your project might be characterized. A way of thinking about this is to consider projects as being measured in several dimensions. As examples, let me explore two dimensions on which projects can be described: *observation versus innovation*, and *study versus case study*. Another useful dimension, which is more or less self-explanatory, is *theoretical versus practical*; or, in some disciplines, *mathematical versus empirical*. Another is *principled versus pragmatic*. In some cases, it may not make sense to even ask where on a particular dimension the research lies—a literary analysis may be neither theoretical nor practical, for example. But in general your description of your contribution needs to be informed by what kind of research you are doing, and this needs to be communicated to the examiner.

Observation or Innovation?

Several of the research topics listed above concern description of phenomena in one form or another: for example, the correlation between advertising and diet; masculinity in literature; diet in Camanderra; and even the (presumably highly mathematical) physics-based atomic models of metabolism. These theses can be seen as based on *observation*, in that they propose a description of observed phenomena and then proceed to evaluate whether the description is valid. In these theses, the description is, or forms part of, the hypothesis.

Other research topics concern construction or invention: a storage plastic, a salmonella test, an energy accountancy method, or a health message design. These theses can be seen as based on *innovation*, in that they involve creation of something new as well as evaluation of it. The description of the innovation may, in some cases, consume a substantial fraction of the thesis. There also needs to be a discussion of the properties that the innovation is expected to have (that is, what is it predicted to do?) and of criteria that it is intended to meet. These components together form a hypothesis, but one that is very different in form to that in an observational thesis.

Other topics are neither one nor the other, such as the food wastage model. A model is by its nature descriptive, but may involve innovative mathematical constructs. Indeed, it is a simplification to regard any topic as purely observational or innovative—no topic is truly at these extremes—but it is a handy simplification, and helps us to understand how different thesis structures arise.

Study or Case Study?

A particular phenomenon can be studied in its own right or to provide information on a broad range of similar phenomena. For example, we might study the farming, processing, and distribution of potatoes because we wish to determine how to better manage particular aspects of bringing this vegetable to the consumer. In this example, there is a clear focus on potatoes: because they are important to the food industry, we need to fully understand the potato production process.

Alternatively, we might study potatoes to determine ways that vegetables are seen in the fast food industry. If we discovered key success factors in the way potatoes are distributed, processed, and consumed, we might then be bold enough to suggest ways that other vegetables could be assimilated in the fast food industry. In the second case, we have used the study of potatoes as a 'case study' to illustrate how vegetables can be assimilated into the fast food industry.

As you can see, despite studying a similar subject in both of the research projects, the structure of the thesis depends in part of your focal purpose and final goal. From the start, then, you need to be clear whether you are investigating a phenomenon *in its own right* or as a *case study* from which you might later develop principles that apply to similar settings. In general, the case-study investigation is far more ambitious and involves more work.

Let me give a more realistic example. One of my students, Hisako, had visited Vanuatu on holidays and was struck by the range of imported foods at a grocery store in the capital. Vanuatu is a developing country that spreads out across a rich archipelago in the south Pacific Ocean. For millennia, human habitation has been sustained by a rich and healthy diet that consists of yams, seafood and a variety of small game. In recent years, however, increased globalization and Western influences have begun to influence the diets of the peoples of Vanuatu. Hisako was interested in the effects that such changes in eating might have on the overall health of the population.

From the start of her thesis, in many of our discussions we sought to determine whether she was intending to write a study, or a case study. We sketched out difference in the two structures and discussed them. The following comparative lists set out differences in report structure, using Vanuatu as an example.

A study, with a focus on a specific phenomenon	*A case study*, as an example *of larger phenomena*
Title: The changing diet of Vanuatu	*Title:* Influences of globalization on traditional diets: A case study of Vanuatu
Problem: As a result of globalization, the culture of Vanuatu is changing, and, importantly, so is the diet. Little is known, however, how global changes are influencing the traditional diet of Vanuatu	*Problem:* Little research to date examines the impact of the increased global production and distribution of food on the diets of people in developing countries
Aim: To examine changes in diet of the people of Vanuatu in an era of rapid globalization	*Aim:* To examine the influence of global trends of food distribution and consumption on traditional diets
Scope: Limit the study to a specific area of the country, potentially just one section of one of the islands, and set limits on the length and extent of data gather procedures by time, participants, and focal aspects of diet	*Scope:* Limit the study to a select range of developing countries, such as those in the Pacific Islands, and note that Vanuatu as a case incorporates many of the characteristics of such countries

Background: Set out the history and culture of Vanuatu with a focus on diet. Demonstrate an understanding of the traditional diet in Vanuatu. Illuminate ways in which globalization is influencing life in Vanuatu with a focus on diet. Create and justify research questions	*Background:* Trace the development of the global food industry. Set out known impacts of dietary change in a range of developing countries. Identify key points of debate where issues arise, and are increasing, in an effort to pinpoint significant areas of concern. Create and justify research questions
Study design: Determine, and demonstrate understanding of, appropriate data collection and techniques. Adapt, for example, an existing questionnaire pertaining to dietary choices. Gather data in Vanuatu	*Study design:* Select case-study approach. Justify the choice of Vanuatu as an example of global issues. Describe Vanuatu. Determine, and demonstrate understanding of, appropriate data collection techniques. Gather data in Vanuatu
Results: Analyze the current choices of food that make up a diet in Vanuatu	*Results:* Analyze the current choices of food that make up a diet in Vanuatu as they relate to global factors
Discussion, conclusions: Within the context of historical records of the Vanuatu diets, set out the results to explain choice in modern diets of Vanuatu	*Discussion, conclusions:* Set out the results in the context of global influences, with a clear link to the theoretical framework and the ability to transfer, or generalize, findings to similar developing countries
Further research; career goals: Propose that a mixed methods approach be used to survey more people (breadth) and understand choices (depth); work for an agency concerned specifically with Vanuatu	*Further research; career goals:* Transfer the study design to a range of developing countries, and nominate key variables to create a survey instrument; work as a consultant to an international organization

The two structures are clearly very different, and the common elements appear at different points. For example, the word 'Vanuatu' does not appear in the title or the aim in the case-study approach, and the description of the Vanuatu case study is deferred for several chapters. Not surprisingly, the discussion and conclusions take quite different directions.

You must, from the start, be clear about which of the two approaches you are using. If you are undecided, you will jump from one structure to the other as the research develops. With regards to the example above, Hisako needed to know right away whether to look for references to do only with Vanuatu or for work to do with a range of South Pacific countries. A discussion of results can be a mess if you sometimes see their relevance in terms of a specific situation, and at other times see the results as an example of larger phenomenon. I offer a test: if you mention the words 'case study' in your thesis, you shouldn't mention the specific area or topic of the case study in your aim or title. If you find that it keeps creeping back into working versions of your aim or title, you have not yet sorted out this problem.

The comparison above demonstrates that there is a leap of faith in the discussion and conclusions sections of the case-study approach, in that it is assumed that the findings for the case study can be generalized. (If you don't go on to at least some generalization, or transference to similar settings, then it is not a case study, but merely a study.) You will have done your best to cover this point in your method section, in which you try to choose the most representative case-study area. However, it may be that you spend so much time developing ideas around your case-study area that you will have little time for the even more important task of seeing how far you can generalize the results to their broader implications. This doubt, over whether the results can be generalized, can be resolved only by checking your conclusions on several other areas, either by doing further work yourself, which you may not have time to do, or by finding reports of comparable work in the literature. So you need to keep in mind that most case-study investigations leave many unanswered questions and pose many hypotheses for further research.

Thus, a case study is, in a sense, a preliminary investigation that seeks to establish an agenda for further research. Yet in the structure shown the case study is the main investigation. This is entirely reasonable. Many PhD theses that are taken up principally by case-study investigations have impressed the examiners. However, I advise you to express the appropriate reservations about the degree to which the findings in your case study are generalizable, and to point out the need for further work to confirm your conclusions.

Method

If you examine a range of theses from the library you will probably see that in many of them the chapter I am discussing is titled 'Method' or 'Research Method'. You may also see 'Methodology' or 'Research Methodology'. But this chapter should include much more than the selection and description of your method. The methods we select are ways of testing hypotheses or answering questions or evaluating innovations. Therefore, if you call the chapter 'Research Method', you are in danger of forgetting to deal adequately with the identification of hypotheses or questions; and if your thesis concerns an innovation, you need to describe it somewhere!

In passing, I have just touched on the need to distinguish between *method* and *methodology*. Often researchers use the word 'methodology' when they mean 'method', perhaps because it sounds more learned. However, methodology is the branch of knowledge that deals with method and its application in a particular field of study. For this reason, social scientists use the word to describe the general stance that the researcher is taking: for example, the researcher as the designer of empirical experiments, the researcher as objective observer, or the researcher as participant in the activities under study. You should reserve it for this usage, and should not use it to describe the design of empirical experiments or objective observations of physical or biological systems. When scientists are designing work of this kind they take a highly standardized stance, the researcher as dispassionate outsider, and don't see a need to discuss their methodology. However, in many areas of social sciences and humanities it is important to tell the reader what stance you are taking, and why. This should be discussed in a separate section on methodology.

In some disciplines, you need to identify your particular stance in regards to the information, or data, that you are analyzing. Do you see, for example, that your interpretation of the findings is best viewed through the lens of a neo-Marxist or post-modernist perspective? Are you undertaking to construct a 'feminist reading' of the incidences that you have witnessed? Will your analysis be coloured by a 'liberal' or a 'conservative' understanding of the political landscape? If you do take a particular philosophical stance towards your study, my advice here is that you explicitly state your views to the reader before you choose which 'method' you will use to gather data. Justify why you decided to use such a perspective from among the many competing viewpoints, and how that perspective informs your choice of data collection methods or instruments. Later, of course, you will need to revisit this view as you analyze your data and discuss your findings.

You have told your readers about your hypotheses or questions. Now you must tell them what method or methods you used to test the hypotheses or answer the questions, and why you chose them. You should first review the methods available to you, and then present reasons for selecting the methods you used. Students often forget this step altogether when writing their theses. You may have used a fairly standard method used by your predecessors for testing the type of hypothesis you have put forward. You may have adopted a method suggested by colleagues or supervisors as being suitable. In these cases you might not really be aware that you had selected a method, but nonetheless a selection has taken place; one of the less obvious aspects of the progress of research is that, not only does knowledge advance, but so does method. There are continual refinements to the statistics that are used to assess the outcomes of experiments, for example.

Alternatively, you might have put a lot of thought into the selection of your method, but by the time you came to report the results, you were so immersed in them that you completely overlooked the necessity to say why you chose that particular method. But whatever the selection process, the reader cannot read your mind. No examiner is going to be kind enough to write, 'Well, even though it is all very unclear, I am sure that the candidate had good reasons for selecting that particular method'—examiners are specifically asked to check whether the methods you have adopted are appropriate, and whether you have justified your selection of them. The issue here is really one of the need to justify your assumptions, which I further examine later in this chapter.

It's worth noting again that, in some disciplines, what I've loosely referred to here as 'the hypothesis' may be a complex bundle consisting of an innovation, discussion of anticipated properties of the innovation, and explanation of criteria that the innovation is intended to meet. 'The method' is the mechanism for evaluating whether the innovation successfully meets these criteria. The examiner needs to be persuaded of the rationale underlying all of these elements.

The term *triangulation* is used in research work when we use more than one research method to answer our research questions or test our hypotheses. The term is derived by analogy from surveying, where precise measurement of something involves observing it from multiple angles or locations, and also ensuring that the measurements are consistent with each other. We might do this if we had more than one hypothesis or question, or if the question was multi-faceted and different methods were needed to throw light on the different facets. This is quite common in research in the social sciences. In other disciplines, triangulation might be used when, for example, the same phenomenon might cause multiple effects, and different methods are required to measure each of them. Consider the example of marsupial dietary adaptation. It should affect relative populations of different species and the abundance of different food sources—both can be used to assess impact. Also, note that both of these are subject to confounds (many things affect population and abundance) that your thesis needs to address.

Use of multiple methods can add great strength to your arguments; how much more convincing it is to have many voices rather than one telling us the same thing. If you use more than one method, though, you not only have to describe each of the methods and why you selected them, but also why one method was not enough. That is, you need to consider the relationship between the methods—to what extent are they independent of each other, for example?

I touched above on the concept of *confound*. Lack of consideration of possible confounds—other reasonable explanations for the same observations—is one of the commonest weaknesses in theses. Your arguments can only be strong if you actively seek confounds, and show that they do not invalidate your results.

'Research Methods'

There is a large literature on the topic of 'research methods', much of it specific to particular disciplines. This literature is largely concerned with the practice of research, or, in the terms I've used in this book, developing the research questions and hypotheses and designing experiments, instruments, and processes for testing these hypotheses. There is also a substantial literature—reaching back many hundreds of years—on the nature of research methods and the philosophical attempt to link observations, experiments, and knowledge. Both of these perspectives on research are beyond the scope of this book.

However, there are respects in which the issues of research methods are intertwined with the development of your thesis. One of them I have already discussed at length: the fact that your research should be shaped, right from the start of your candidature, by your writing. Until you have made the attempt to capture your hypothesis, experimental design, and so on in a precise form, you aren't really ready to explain and argue for your research program. That is, you need to be in the position of being able to say, as unambiguously as possible, that '*this* is what I am doing' and '*this* is how I am going to investigate it'. Only once you have a concrete *this* can your colleagues and supervisors debate it with you and help you to sharpen your arguments and your thinking.

Another respect in which method and thesis are intertwined is a subtle point that many researchers overlook: the issue of measurement, or assessment. A researcher who uses thousands of words or more of careful argument to explain a hypothesis and its importance, and thousands more explaining how it is to be evaluated, may completely fail to justify the method being used to measure the outcomes.

Let me give an example. Suppose your project concerns food wastage due to the limitations of distribution networks, such as, for example, the loss that occurs when a shipping container of vegetables spends an extra day in a warehouse. This can be caused by factors such as: the vegetables should not have been shipped there in the first place; there is no truck or driver to distribute the vegetables after shipping; there is no capacity to receive the vegetables at the retail outlets. You have pro-posed a mathematical model that (in principle) a grocery chain could use to try and reduce such loss, by, for example, integrating driver rosters with data extrapolated from recent sales. As a measure, you use the tonnage of vegetables that are thrown away due to their having reached a sell-by date; that is, you deem your model to be a success if it can predict a schedule that reduces the weight of vegetables that are discarded.

This measure sounds plausible, and is emotionally appealing because the idea of throwing away tons of vegetables seems such a waste. But do you think that it is sufficient to simply use 'tonnage discarded' without some rationale, or with-out some consideration of the alternatives? I hope that you agree with me that the answer is no! It certainly isn't good enough to use a measure on the grounds that it has some emotive impact. To see what I mean, consider some of the alternative measures (not all of which you will agree with, but I think you will see that some kind of argument could be made for each of them): the retail value of the discarded vegetables; the energy saved by not delivering vegetables that won't be sold; or the value of the vegetables that can be discarded at point of production instead of point of sale (allowing them to be used, say, to feed pigs or as compost, rather than landfill). To take the first of these, an economist could argue that it is reasonable to discard cheaply grown, cheaply harvested local vegetables in preference to expen-sive vegetables that have been shipped over a distance, even if this increases the total tonnage thrown away.

Even if it is 'obvious' (a loaded term! If it is obvious, you should be able to briefly explain why) that your measure is correct, it then needs to be applied to a concrete problem, such as a particular distribution scenario. In this scenario, you might postulate a mix of vegetables sourced from local and remote farms with a mix of lifetimes and of shipping and storage environments (chilled, humid, frozen, and so on), and a mix of local distribution mechanisms and kinds of retail outlet. The richer this scenario, the more realistic it might be—and such research would seem to be of limited value if it only applies to idealized scenarios—but the more assumptions it embodies. This is a version of the study-versus-case-study problem; the more specific the concrete problem, the greater the leap of faith required to be-lieve that the solution will generalize to other cases. To address this issue, you might need to apply your model to hundreds of scenarios representing different mixes of the various factors.

Something I've found intriguing is how easily, under prompting, some students find problems with the measures they have been using. A particular example was a Masters student I was working with, Vivienne, who was looking at the effectiveness of different forms of the 'traffic light' labelling that is used to indicate the healthi-ness of retail food. For each of the measures she had proposed, and even one she had worked with for several months, she quickly found reasons why it might be unreliable. (These included measures such as shifts in sales volumes and survey questions to consumers. You might want to think for yourself about what the shortcomings might be.) It was as if it had never occurred to her to think about whether each measure was plausible or not, but, when faced with a direct question, she could readily find reasons to query them. That is, she hadn't considered whether there were confounds that would undermine her method.

Another perspective on the issue of measurement is that, often, the researcher has made assumptions that other researchers would not see as justified. A common problem that I see is that measures are just too simplistic—in the food wastage case above, obviously the total tonnage is a concern, but it is also a concern if things with a low production cost are kept while things with a high production cost are discarded. You need to try and expose your assumptions, and justify them. As for your hypothesis, you need to write about the assumptions underlying your measures in order to tighten them into a form where they can be debated with others. It may be that the measures themselves are something you have assumed, and not made explicit; it is critical that you directly ask yourself how the outcomes will be assessed, and what the measures are, so that these decisions are made clear to the examiner.

The third interaction between method and thesis writing is your need to be confident that the evidence you are gathering will in fact test your hypothesis. It may seem surprising, but I have often seen research projects in which this is not the case. Quite simply, somehow it happens that people start gathering data that isn't going to help them. I suspect that there is no general cause for why this happens. It may be, for example, that a researcher goes looking for some particular data (say, of numbers of people participating in a particular sport) but finds that it is inconsistent or unavailable, and instead decides to make use of other data (say, of numbers of people who attend sporting events) that seems as though it will be a good proxy; and then some other level of approximation or indirection is introduced; and all too soon the connection to the original argument is lost. Or it may be that the project was inspired by the appearance of some particular data source (the discovery, say, of historical records of goods shipped to Australia on convict vessels) but that the hypothesis that was to make use of this data had, through the course of discussion with the supervisor, shifted away from issues that the data could resolve.

Thus it is critical that you regularly assess the relevance of your data and method to the aims of the project. If you have been following my advice, you will have written down the aims as part of your introduction; you need to check these aims against your method to make sure that they continue to be consistent with each other.

Argument

The issues of measurement and relevance are aspects of the need for your conclusions to be built on a robust *argument*. As I discuss in Chap. 8, you will use your data and method as part of a path from raw results and numbers to new knowledge;

this knowledge is the primary purpose of your thesis. This path, or chain of reasoning, is the argument.

In the context of your method, the main thing is that you need to be confident in at least a preliminary way that the path exists. That is, you need to satisfy yourself that if you gather certain data, and it has certain desirable properties, then you can use it to present an argument that your hypothesis probably holds; or that, if the properties are absent, then the hypothesis is probably wrong.

To continue the earlier example, it might be that your hypothesis is that complex modelling can reduce vegetable wastage, and applying your model shows that it can reduce wastage on a certain scenario by 15%. The argument first needs to persuade the reader that 15% is a meaningful reduction: is this a significant percentage of the total volume of shipped vegetables? What are the costs incurred in the reduction? And so on. If you are being intellectually honest, you may have, early in your research, considered the question of thresholds; for example, perhaps no case can be made for the benefit of a mere 3% gain, while a 10% gain is unarguably significant, and so 15% is a clear success for your model.

The argument should not stop here, because so far it is rather inward-looking, with results that show that the research is valid in its own terms, but only weak evidence that it has external value. To take the argument further, you might then consider the issue of generalization to other scenarios, and also even consider broader factors, such as demonstrating that the data that the model needs can be gathered in practical settings.

A perspective on a thesis is that it is a presentation of an argument largely made in the results and analysis chapters, but its foundations are laid here, in the chapter on your contribution. At this point you need to have established for yourself what the lines of the argument are going to be, and have addressed likely sceptical concerns the examiners might have by examining and justifying your assumptions.

A final note on this point, but a significant one, is that ultimately the argument and your assumptions are subjective. Your final line of defence, in choice of assumptions, method, and so on, is that they are *reasonable* and *consistent*. The final product of your effort is an objective thesis, but it would be dishonest to disguise the fact that some elements are essentially choices, whether they arise from constraints (only certain data was available, resources were limited, and so on) or from explicit decisions (inclusion of energy considerations made the model intractable). At the same time, you need to be confident that the reasonableness of these choices is obvious to others. If you and your examiners cannot agree on the basis of your research, they are unlikely to respect your outcomes.

Organization

So where do these lines of discussion take you in terms of how to organize the central chapters of your thesis? As I noted earlier, practice varies a great deal between disciplines, and you need to read other theses in your area—and of course get advice from your supervisor—to establish how your thesis should be arranged. For some topics it may be, for example, that this part is quite brief, with a focus on statement of a hypothesis and an explanation of how data was collected in the process of evaluating the hypothesis. For other topics, it may be that this part extends over two or more extensive chapters, which will contain descriptions of an innovation, a discussion of what is involved in practical deployment of the innovation, explanation of criteria the innovation needs to meet, and a description of the experiments that have been to used to evaluate the innovation. The practice of your discipline is the best guide, and if you have followed my advice and sought out other theses you will undoubtably have found good examples to use as models.

But while I cannot give you guidance that is specific to your discipline, I can give you advice on the criteria that this part of the thesis needs to satisfy. The term is overused, but let me again say 'narrative'. In this part of the thesis more than any other, you are leading the reader through your thinking, and need to do so in a way that lets the reader feel that your hypotheses and methods are reasonable and appropriate. You need to explain why your proposals are plausible, and at least intuitively offer advantages compared to other perspectives or approaches.

- For an observation-based thesis, this may flow directly from the background material; for an innovation-based thesis, this may involve, say, your building a case that your new approach solves problems that previous approaches neglected.
- For a study, you need to persuade the reader that the subject is of sufficient interest; for a case study, you need to persuade the reader that the subject is representative of a broader population.
- A quantitative thesis may need sections, or a whole chapter, on experimental design and data collection.
- If you have used triangulation, your narrative needs to introduce the need for multiple methods, then describe them and explain how they support each other. You may need to have multiple separate sequences of presentations of methods, results, and analysis.

And so on. Each project will be different.

A common theme, though, is that you have told the reader what research method you used and why you chose it. Before you describe the results obtained by using this method, you must first describe in detail the *way* that you applied the method, and *why*. Although projects may use quite different methods, the points to be dealt with are similar: clear identification of hypotheses; explicit choice of method; design of research instruments to test hypotheses. With these in place, you can proceed to the presentation of your results.

Summary of Chapter 7: Establishing Your Contribution

Positioning your work:

- Draw on the conclusions of the background chapters to identify your research hypotheses or research questions.

- In the social sciences or humanities it may be necessary to describe both the research program and the stance you have adopted as researcher.
- Understand where your work lies on the dimensions that are appropriate to your discipline: theoretical or applied, study or case study, and so on. Use this understanding to inform your explanation of your intended contribution.
- Positioning requires justification. For example, if you decide on a case-study approach, explain why, and justify your choice of specific case to study; if a long description of the case is needed, note that you will do this in a separate following chapter.

Method:

- Discuss the range of research methods that could be used to test your hypotheses or answer your questions, and choose the most appropriate. Don't forget to justify your choice, even if it is standard for your discipline.
- Experimental validation requires an experimental design. Expect to have to explain it in detail, and also to justify it. Make appropriate choices of measurement or assessment mechanisms.
- 'Method' and 'methodology' are not the same thing. 'Method' refers to specific techniques and 'methodology' refers to the stance you are taking as the researcher.
- If only one method is to be used, describe the research instruments to be used in implementing it. If more than one method is to be used it is usually better to defer the descriptions of research instruments to the particular chapters where you implement the methods and obtain results.
- If you have not already described your detailed research procedure in the 'Research Design' chapter you should describe it first before you go on to report any research results.

Organization:

- Design a narrative flow that takes the reader painlessly through the central part of your thesis—the part that consists of the new ideas that you are arguing for (and is thus the most unfamiliar).
- Be alert to the different narrative structures used in different kinds of thesis.
- Ensure that at all stages you have a clear understanding of the argument you intend to use for linking of question, data, analysis, and outcome.

Chapter 8
Outcomes and Results

In a typical thesis (or research paper), data and argument are used to build a case. That is, a logical narrative is used to persuade the reader that the claims of the thesis are reasonable and are supported by evidence. From this perspective, maybe half of a thesis can be viewed as a sequence of three components: first, how the data was gathered and what it is intended to represent; second, what the gathered data looks like; third, how it should be interpreted. How to present 'what the gathered data looks like' is the subject of this chapter.[1]

If you have been undertaking quantitative work—bench experiments, surveys, measurements, and so on—clearly you will need to report the outcomes of your investigations. What should you include in the 'results' chapters, and what should you leave out? At this stage of the research, you will have analyzed and interpreted your results, and now you need to use them to present an argument to the reader. If your work is more qualitative—case studies or reviews, for example—you probably still need to present an objective review of what you have found, and it may well be in the form of a 'results' chapter. In either quantitative or qualitative work, such a chapter provides a basis for the analysis or discussion that completes the body of your thesis.

It is true that some theses don't have a results chapter; my (Zobel's) thesis, for example, primarily consisted of a series of linked mathematical results, in which

---

[1] A note on terminology. Discussion of how to present results is clouded by the inconsistencies in the way experiments and their outcomes are described. In many fields of research, for example, *data* is the outcome of the recording of measurements. The data could have been recorded by you as the researcher using the instruments you devised to test your hypotheses, or recorded by some other researcher and then made available. Or they could have been recorded for some other purpose, such as the temperatures recorded at a meteorological station, or the share prices recorded at a stock exchange. But data can also be the subject of an experiment. A researcher investigating a weather model could use temperature measurements as an input, and the recorded values—'data' in the above definition—could be the input to the model, which also produces 'data' as output. Here I use *data* to describe experimental results, or measurements, and *outcomes* or *results* to describe what the researcher found by interpreting these measurements.

D. Evans et al., *How to Write a Better Thesis,* DOI 10.1007/978-3-319-04286-2_8,
© Springer International Publishing Switzerland 2014

each of the main chapters was built around a theorem and a proof. Another instance is a thesis in which the student used three case studies (her own patients) to explore the relationship between childhood neglect, high social status, and psychopathy; a chapter was dedicated to each of the case studies, which led straight into the final discussion. A third instance was a historical study weighing up the evidence for and against water contamination as the real cause of a 'plague' in the 1600s, where much of the challenge of the thesis was the fact that the data was sparse and unreliable. However, the 'how to present results' thinking discussed below can be as relevant to such work as it is to a thesis with a more quantitative organization.

Quantitative or Qualitative Data?

A common categorization of research is that it is either *quantitative* or *qualitative*, or perhaps more accurately, whether it is closer to one or another of these extremes. The experiences of two students I worked with are good examples of the challenges in quantitative work. One, Jorge, had had the luxury of being able to test his idea (a way of reducing the time required to compute some kinds of simulations) over a great many data sets. He had small sets he had used for preliminary measurements and much larger sets for the final evaluation. However, the data was not always consistent, and statistical evaluation of the data, and visualization to confirm his understanding of the statistics, had played a part in forming conclusions. As a consequence Jorge had some hundreds or more of graphs and tables to draw on, reflecting tens of thousands of automated experiments (and he could easily have run many times that number), and now he needed to use this material to construct a narrative. Despite the strong quantitative basis, however, the data fell into cases that needed qualitative analysis.

The other student, Dai, had painstakingly gathered a small set of data through work with genetics researchers, tracking the extent to which a software tool for comparing chemical structures was reliable enough for practical use; this data was counts of effects such as the number of incorrect comparisons, and human estimates of factors such as the severity of the error or the extent to which it caused the researchers to waste their effort. Here the challenge was that the results were to some extent preliminary, but nonetheless strongly indicative; they went beyond a simple case study, but didn't easily fit into a tabular or graphical presentation, and was insufficient for thorough quantitative analysis.

The task faced by both Jorge and Dai was to take a chaotic mass of information and turn it into a results chapter. In both cases it was reasonably clear that the results supported their initial hypothesis, and the challenge was coherent presentation. For another student, Jackie, there was a deeper challenge because the data was contradictory. She had come to me through a mutual friend because, as she assembled her results, she increasingly wondered if her interpretation of the results was strong enough to stand scrutiny.

Jackie's work was about food choices and its relationship to secondary education outcomes. Put simply, did what teenagers eat affect how well they went at school? The problem for Jackie was that the topic is part of a complex network of societal, economic, and political issues. There was almost an overabundance of data (she was reinterpreting materials previously collected by others), which had been previously harnessed in conflicting ways by a range of people and organizations. Many of them had used it selectively to support points of view that were driven by ideological goals: for and against providing meals in schools; for and against imposing dietary guidelines on school canteens; blaming parents for neglect; blaming the state for intervention, or failure to intervene; and so on. Worse, some of the data collection had been undertaken for political ends, and so was suspect.

This data had both qualitative and quantitative aspects, but it was the quantitative aspects that were relevant to Jackie, who was trying to identify whether there were some biological underpinnings to the problem, and in particular had formed the view that chronic lack of exercise rather than specific diet was the more identifiable culprit in poor academic results. However, she had gotten lost in the detail, and had begun to worry about whether she might get publicly criticized by special interest groups if her findings were made public.

A more qualitative example is that of a student, Don, who was examining food imagery in classic novels. He had collected forty or so examples from separate literary traditions that he felt supported his theme on how 'meal episodes' and food imagery were used to establish the morality (or otherwise) of the characters. For example, in some novels the first time the reader discovers that a character is selfish is when the character doesn't share food. Don had already written detailed case studies that supported his argument in the context of several novels, forming Chaps. 3, 4 and 5 of his thesis, following a thorough review of relevant aspects of literary theory in Chap. 2. He now wanted to use his wider collection of examples to demonstrate that his argument held generally, but was struggling to assemble the material into a persuasive form.

Despite the superficial differences, in many respects these students faced much the same challenges, as I now explain.

From Data to Results

You have a hypothesis; you have been busily gathering data and drawing inferences, and, informally at least, linking the data to your original goal. Now you have to take this activity and use it to persuade the reader to agree with your thinking. The process starts with your data.

What 'the Data' Is Comprised of

One thing that these examples illustrate is a critical but hidden issue: each of these students had been making decisions as to what body of material they were thinking of as 'the data', and were managing that data somehow. That is, before you can begin analyzing your data, you have to have some data to analyze.

A first step is to decide what is 'in' and what is 'out'. This is perhaps most obviously an issue in Don's case—what actually constitutes a 'meal episode'? What is the basis for choosing novels to analyze? What literary traditions should be sampled? These were issues that he had tackled in his early chapters.[2] Once the inclusion criteria were settled, and he was reasonably sure he had a representative selection of cases, he could then proceed to use the cases to develop his findings.

A next step is to systematically organize the data. I cannot emphasize this enough. You need to create spreadsheets in which the data is laid out in a regular way, or build files in which material has been categorized by key criteria, or draw pictures showing how the data items relate to each other, or something else; but whatever you do, get the data under control. When a student walks into my office clutching a big pile of scruffy printouts, or shows me a Windows folder full of files with no idea of what the file names mean or what is in them, or has lost track of which version of the data is correct, or which graph is current, I know the student is in trouble. A casual approach to managing your data may not seem to create issues early on, but leave things too long and the complexities will compound and soon get out of control.

Materials that are in a mess suggest that the thinking is in a mess. This is a good point for self-reflection: if you find that your arrangement of the materials has become chaotic, then maybe your grasp is chaotic too. Take yourself back to first principles, ask basic questions about the data and what it is supposed to represent, think about how you would like to see it organized—and then make it happen. Remember that a core skill of research is careful thinking. Take heed of signs that suggest that you can improve, and act on them.

A good presentation of results rests on having the data—which as I noted above may be voluminous and contradictory—organized and under control. And this, in turn, rests on clear principles for what the data is: what is valid, what is included or excluded, and so on. The lesson for your thesis is that the reader needs to know too. An examiner won't trust your results unless they understand that your data is fair. A clear presentation of how the data was chosen, what its properties are, and so on, is essential to establishing trust with the reader, and, just as importantly, satisfying yourself that your data is complete and correct.

---

[2] When faced with doing the analysis, Don began to made some fresh judgments as to whether something should be included; an obvious risk in making judgments during the analysis is that subconsciously he might exclude cases because they did not fit his hypothesis, and initially he did make errors of this kind. What it also reflected was that his thinking had developed, and he needed to adjust his early chapters accordingly.

Presentation

In the previous chapters of your thesis you described the design of your work, explaining how it tested your hypotheses or answered your research questions. You now have to present the results you obtained in this work.

This presentation should not be haphazard. The presentation should *educate* the reader. You may believe that your task is to include every single data point or case that you recorded in your work—but doing so is almost certainly a mistake. You have used this data to draw conclusions as objectively as you can; now the task is to use representative examples drawn from the data, and example analyses of the data, to persuade the reader of the validity of these conclusions.

In any case, even when the data is limited it is surprisingly difficult to capture it all within the confines of a thesis. In Don's case, even a brief explanation of a single 'meal episode' might take a page or two; in Dai's case, a single transcript of how his method was used in the course of a study of a chemical structure might take ten or more pages. Jorge's raw results had millions of individual data points, and thousands of secondary products could be built on these, such as tables and graphs showing the cost of his simulation method under different assumptions. Inclusion of all the data is unlikely to be feasible.

And what would be the point of simply dumping the data into the thesis? It is unlikely to be meaningful to the reader. Here are the things the reader needs to know, some of which may have been covered in earlier chapters:

- How the data was gathered—where it was sourced from, what aspects of it were measured, what it consists of, what the guidelines were, what permissions were required, what restrictions apply, and so on.
- How the data might be obtained by a reader—whether directly from you, or from what external source; or how similar data might be created.
- What the results looks like—by example; or by graph, to show, say, the distribution of values; or by tables of typical instances. For example, a common strategy is to list out the categories into which the data can be placed, and give an example of an item in each category.
- Summaries of the complete set of results, in as rich a way as possible.
- Notes of issues such as known gaps or incompleteness in the results, or where the data may be uncertain or unreliable.
- Analyses of the results, using discussion, argument, statistical tools, and so on, as appropriate to the work.
- Interpretation of the analyses, completing a transformation from data to knowledge (more on this later).

Inclusion of the raw data is not in this list. Consider Tony's work on laser grading mentioned in Chap. 6. This included focused interviews, lasting about one hour each, with three farmers and several professionals in the area of land management. All were recorded and transcripts of them prepared. Should Tony have spent so much time preparing transcripts? Having prepared them, should he have included them in his thesis? If so, where—in this chapter, or relegated to an appendix? By now, you should be able to answer these questions. My response would be to include enough of the data (probably in an appendix) for the reader to see how it was collected, what form it took, and how it was treated in the process of condensing it for analysis. It would makes sense in this case to include in the appendix the transcript of one focused interview; the others would be kept on file for reference.

Note by the way that many disciplines, and most institutions, have research guidelines that concern how data is managed and described. Make sure you are familiar with these guidelines as you work through your data presentation.

Another example is the work of Geoff. In his study of the ignition of brown coal particles, he had tested 96 combinations of experimental conditions, with 10 particles tested at each combination, giving results for 960 particles in all. First he averaged the ignition times of the group of 10 particles, and gave the standard deviation to indicate the variability, thus cutting down the entries in the table from 960 to 96. He presented all the results for 1000 micron particles first, then 500 microns, and so on. He then arranged the results for each particle size in groups for each of the other variables. The reader could then examine the results with the hypotheses in mind, and develop mental pictures of the effects of the different variables. Geoff could have gone on to plot some of these effects as graphs, but he preferred to wait until the next chapter to do this, because he wished to plot the results against what he would have expected from a theoretical model he had developed earlier in the thesis.

Somewhere you may want to discuss the data you didn't gather, but would have liked to, and other issues of that kind. My advice is that this is a good thing to do. No piece of research is really complete, and the reader will appreciate your views on where the work could be extended. For the examiner, such discussion shows that you are thoughtful enough to be aware of the shortcomings of your study. Remember that the examiner may well notice these shortcomings without your help—and acknowledgment of shortcomings is not a sign of weakness. Look again at the criteria for examination of PhD theses, and note that 'an awareness of limitations' is expected.

Being blind to shortcomings is not a characteristic of an effective researcher. An example is my student Kirk, who was always keen to move on to the next experiment before the current one was complete, and—as he knew, but would not openly admit—he was not as capable in the laboratory as other students. However, he was good at presenting his work, glossing over flaws and limitations. While his work was never precisely fraudulent, his write-ups (for his thesis and for research papers) would avoid the weaknesses and emphasize the strengths; a problem that eventually brought on a crisis, when he had to repeat some work and no longer observed the effect he had originally claimed.[3] It was a straightforward case of carelessness,

---

[3] Almost certainly because of one of the classic fallacies of science. If you repeat an experiment with small variations until it succeeds, and then stop, it may seem as though the desired outcome is achieved; if there is any kind of uncertainty in the experiment, it is just as likely that the positive outcome occurred by chance. If this success is all that is reported, the reader gets a highly distorted view of the true results. This is known as confirmation bias.

haste, and overconfidence—and a lack of willingness to listen to criticism—leading to difficulties later on.

Analysis

Two key concepts in every aspect of managing data and presenting results, which I have touched on a couple of times in this chapter, are *variables* (or *parameters*) and *category*. These concepts reflect our understanding of the data. We want to understand what kind of data we have—what sort of 'meal episode', for example. Assigning instances to categories lets us discuss and analyze data in a consolidated way. Variables determine the behaviour of the data, and we have understood what is going on when we can accurately predict how variables and data values interact. These concepts underpin how we proceed with data analysis.

Having presented the data in an informative way, how much further should the analysis go? In the work discussed earlier, Geoff made some general remarks about his observations before he tabulated the results. The chapter stopped abruptly at the end of the tables. He felt he had a good reason for doing this but, as I look back on the work, I find it unsatisfying. He had some strong hypotheses to test. He had designed experiments to test them, and had carried out the experiments. Were the hypotheses upheld or rejected? The reader wants to know what the findings are before the writer goes on to discuss their implications.

And it is not only the reader who is learning. Your presentation of results is part of your process of interpreting them—writing the results chapter is part of a cycle of understanding, not an end point. Your aim is to educate others, but self-learning is likely to be part of the process, even at this late stage of thesis writing.

In complex situations such as the coal example above, in which there is considerable interplay of the effects of the different variables, it may not be easy to disentangle the results and their implications. Nevertheless, I recommend that you try—plot the results in terms of the hypotheses. Geoff could have plotted ignition times (a dependent variable) against the moisture content of the particles (an independent variable), with all other independent variables held constant, or against the size of the particles, or the oxygen content of the gas, or its temperature. He did these plots for his own information; but he decided not to present them as part of the results chapter because he had found that the interpretation of the results was not as straightforward as his original hypotheses indicated. In retrospect I believe the reader would have been in a better position to go on to the discussion chapter had he presented the results as tests of his hypotheses by plotting dependent variables versus independent variables. Then, in a very brief discussion, he could have pointed out the unexpected complexity, and announced that he would be dealing with this in his next chapter.

But don't go to the opposite extreme. In one thesis I examined the candidate had discovered the power of a chart-drawing facility. His results chapter contained over a hundred charts plotted by trawling through all of his data sets and plotting every variable against every other possible variable in an effort to analyze the data. His readers were given so much information, at such a low level, that they were totally overwhelmed, and learnt nothing about the system being investigated. The candidate should have confined himself to plotting charts that tested his hypotheses or that demonstrated something significant.

My dissatisfaction with Geoff's chapter on results was due to its unfinished nature. Hypotheses had been tested, but we did not know what had happened. An aim had been stated at the beginning of the chapter (to report the results of experiments to test hypotheses), but it had been only partially fulfilled.

If you turn back to Chap. 4, in which the structure of chapters is discussed, you can see how important it is that you state the purpose of the chapter in the introduction, and that you write a conclusion in which you describe how that purpose has been fulfilled. This rule is as important for the results chapter as for any other. At the end of the chapter, you should share with your readers your understanding of what is now known that was not known when the chapter began.

Reasoning From Data

A common problem for each of Jorge, Dai, Jackie, and Don was that they had formed views based on their data but had not yet done the detailed work of building an explicit argument that presented what it was in the data that supported these views. The aim is to use your data to make a case for the proposition being explored in your thesis. Consider the dictum: 'data is not information, information is not knowledge, and knowledge is not wisdom'. As soon as you use data to test a hypothesis—that is, make a link between a proposition and some observations—they become *information* (what the data tells us). Temperature measurements collected by the weather bureau becomes information when, say, used in conjunction with records of plant growth to test the hypothesis that plants grow faster at higher temperatures. Similarly, the data collected using your research instruments becomes information when you use it to test the hypotheses that led to design of the instruments.

Information becomes *knowledge* when you use an argument to draw conclusions from it: that plants do grow more vigorously at higher temperatures, for example. Here, the argument—a chain of reasoning—is an explanation of how the information demonstrates the conclusions, and will probably need to explain why other plausible interpretations of the information are likely to be incorrect (that is, you will need to eliminate alternative explanations). The argument will also need to explore the subtleties of the data—to demonstrate that the number of measurements is sufficient for statistical significance, for example.

Knowledge becomes *wisdom* when it is integrated into your whole way of looking at things. It is the implications of the conclusions you draw from your results that become wisdom: new insights, new theory, new frameworks. That is, your results chapter is a keystone of the hypothesis-evidence-argument-theory (HEAT) structure of much research, which can be sketched as:

- Development of a proposition or initial hypothesis, which is used to shape the gathering of some observations.
- Formation of a definite hypothesis.
- Building of tools and use of them to gather measurements to be used as evidence.
- Construction of an argument that uses the evidence to give a case for or against the hypothesis.
- Conclusion by developing a new theory or framework.

(As an aside, too many students—and some supervisors!—confuse theories and hypotheses. Theories are the outcomes of research. They represent our most certain comprehension of the universe: the theory of relativity, the theory of evolution, and so on. They are the things in which we have the greatest confidence.[4] A hypothesis is an unconfirmed supposition. Another, arguably worse, confusion is between theory and speculation; some people think they are theorizing when they propose new untested ideas, but from a more formal perspective they may be doing little more than guessing. While such sloppiness is fine in conversation, it has no place in a research thesis).

The HEAT analysis of the research process points towards what you should include in the results chapter and what you should leave out. Raw measurements do not convey knowledge unless you explain or display them in a suitable way, and should be left out or just possibly relegated to appendices. Results displayed in the form of tables or figures that enables you and the reader to make sense of it becomes information, and should be included. Having presented the information, and explained how it is linked to the initial hypotheses, you can draw some inferences from an examination of the information. This will include considering the individual sub-hypotheses that you put forward, and proceed to interactions between the variables that you may not have expected and, if you are lucky, to some totally unexpected results.

You and the reader now know something that you did not know before you carried out your own work—you have transformed information into knowledge. At this point, stop. Keep your theorizing about this for your 'Discussion' chapter, for it is there that you advance from knowledge to wisdom. It is the implications of the conclusions you draw from your results that become wisdom: new insights, theory, paradigms.

Quantitative or Qualitative, Revisited

In this chapter I've suggested a framework for an effective presentation of results. This broad framework was used by the four students discussed at the start of this chapter, whose stories I now complete.

---

[4] In some attacks on science, certain kinds of knowledge are condemned as 'only a theory'. People who use such arguments have failed to realize that the mathematical constructs that are used to design planes or transmit electricity are, also, 'only theories'. We may not be certain of their truth, but they are the knowledge for which we have the most consistent evidence.

Jorge's case was relatively straightforward. He had a lot of data, and the tools to build a variety of graphs and tables that interpreted the data in useful ways, including statistical summaries, trends, and behaviour as different variables, were tuned. Having spent months investigating the data, drawing preliminary conclusions, and then seeking confirmation (or confounds), he could then choose typical examples to illustrate his argument. He first listed the data sets and the analyses applied to set context for these examples, and in some cases could then just summarize the outcome of the analysis. By having reasonably objective criteria for choosing what to present, he was able to build a persuasive qualitative description of the work he had undertaken, and how it supported his original hypothesis.

Dai's results chapter was, in the end, similar. He had to be exploratory to find ways to consistently describe the issues encountered by the different scientists he studied, including assigning issues to scores in the range -3 to $+3$ to quantify their severity, and, separately, carefully explaining how he had decided which problems were more important or less important. Once he had identified common themes, and thus a categorization of the cases, the presentation was straightforward. That is, he could consolidate his results into tables, and used them as the basis of a discussion showing how they confirmed his initial hypothesis.

Jackie's problem was that, fundamentally, she didn't trust some of the data, and needed to build her arguments with unusual care. The approach we took was to build the presentation from small units, each of which represented a single logical step and which, we felt, could be defended by a simple, unarguable case. We began, not by looking at the data, but by going back to basics and setting out criteria that a trustworthy data set should satisfy. In particular, we identified potential sources of bias or distortion in the data sets, and also tabulated the kinds of factors that might lead to the results observed in each case. For example, poor academic results might be a consequence of poor diet—but it might be that poor results lead to depression, which then leads to poor diet. Some studies attempted to control for such factors to try and distinguish between these alternatives; other studies were less well designed. One particularly good study noted which students were siblings, so that, for example, by working from the assumption that siblings usually have similar diet then it is possible to explore the degree to which differences in performance are due to other factors.

This foundational analysis allowed Jackie to organize the studies by their strengths and defects, and undertake meta-analysis (that is, combined analysis of a group of studies) on the basis of which criteria each study met. She then worked through the studies re-analyzing the conclusions that had been reached, and showed that some of these conclusions could be refuted. With the field of explanations greatly reduced, she could then examine where in the data her own hypothesis was and wasn't supported, and ultimately was able to reach nuanced conclusions. In contrast to the sometimes dogmatic assertions made in earlier work, her results were thoughtful and reasoned, and she avoided the mistake of making overly strong claims.

For Don, the initial hurdle was that he had not appreciated that he was trying to present what were essentially quantitative results; the approach he had taken earlier in his thesis, where he had explored in detail how each case had linked into the framework of the novel—thus identifying consistent patterns in the cases—would not work across a larger set of examples. My impression was that crossing this hurdle was genuinely a challenge for him, as it changed how he thought about the outcomes of his research. Instead of reasoning about whether something did or didn't happen in a novel, in effect he now had to assess or measure the extent to which it happened. He needed to use the initial cases to identify specific criteria against which subsequent cases could be evaluated. He then worked through a couple of cases to show how he assessed them against these criteria, as examples to show the reader what he was doing.

The next step was to assess all forty cases against the criteria, and report just the assessment; the cases themselves could be sketched in a few lines each (these sketches were in an appendix) with the details recorded in his research notes, but not directly reported. Given this tabulation of the cases, he could then proceed to draw his broader conclusions. However, in doing this he also needed to acknowledge that he had been somewhat selective in choosing the cases, and as a consequence realized he had to weaken his conclusions. As Jackie had done, he shifted away from a somewhat dogmatic position to arguments that were more reflective.

As for so many elements of thesis writing, the best examples are likely to be theses in your research area. I suggest you locate a couple of theses that have results chapters for research that is similar in approach to your own, then use the framework presented here to review these results chapters and identify their strengths and weaknesses. Your task then is to reproduce the strengths and address the weaknesses, to produce a good results chapter for your own work.

Reflection

In each of the four cases above, the process of assembling a results chapter led the students to reflect on their results and to some extent re-interpret them. Perhaps, you might argue, this reflection should have happened earlier. My sense is that it *can't* happen earlier—it is the discipline of writing it all down in a coherent form that allows identification of issues.

On that reasoning, there are two lessons. One is that the writing of this chapter should not be left too late. Indeed, the writing of all of the chapters shouldn't be neglected, but this one is particularly important, because no student wants to be in the position of changing what their thesis is about at the last moment.

The other lesson is that doing research is a continual cycle of activity, learning, reflection, and refinement. It doesn't stop when the method is 'proved' or the results or gathered or even, often, when the thesis is submitted. You should expect to continue to have new insights about your work—and, perhaps unfortunately, to find new problems in it as well—for a long time after the bulk of your investigation is complete. You should always be prepared to look at your work with a critical eye and, if there is a problem to address or a refinement to make, then do so.

Illustrations

A typical results chapter consists of argument and narrative supported by illustrations, that is, graphs, diagrams, pictures, and tables. But why are you using them? The immediate answer to this question is typically: 'I include a figure when it expresses the point I wish to make more clearly than does the written word'. On this principle, illustrations are likely to play a role in many parts of your thesis. I discuss them here because the results chapter is one place where they are not just helpful, but essential.

However, the use of illustrations to make a point is, I believe, only part of the reason they are of value. If you wish to get the best out of your illustrations, you need to put yourself in the position of the reader. Do they read the written text until they get to the sentence, 'Fig. 6.2 shows that increasing the population density decreases the per capita consumption of petrol', and then dutifully find Fig. 6.2 to check that this is indeed so? Probably not. My experience is that, long before readers begin to work through your results chapter, they will have opened the thesis and skimmed through it, 'reading' the diagrams and looking at the graphs and tables. It is this, rather than the careful reading that comes later on, that seeds the examiner's understanding of your results.

After other preliminaries, such as looking for the aim of the research and reading the conclusions, the real reading begins. The written text develops ideas in the way that the writer intended, and readers will no doubt follow this development. But at the same time they will be generating their own interpretations and impressions. They will compare material in one chapter with diagrams or text in another in ways that the writer had not anticipated. They might refer to and puzzle over Fig. 6.2 long before they read the text that discusses it. They might return to it again when something written in Chap. 8 triggers another train of thought. Such exploratory reading is an essential part of thorough understanding of complex work.

Readers use several complementary channels of communication simultaneously, some using words and some using visual images. They do not use one at a time, switching from one to the other; rather, they use all of them at the same time, perhaps giving one more attention than others at any given moment. Nor is learning always linear. Think of lectures you have been to where the lecturer has used slides to complement the spoken word. You are busy looking at one of the slides and thinking about it, while still listening to what is being said, when suddenly, much to your annoyance, the slide disappears. The lecturer, already busy with the next point, didn't think it was of any more interest to you, although you were busy integrating it with the rest of what was going on in the lecture.

This leads to some rules about visual material:

- The reader should not have to read the text that refers to the illustration to understand what the illustration is meant to demonstrate. Although an illustration should always be 'called up' by the written text, it should make sense by itself. In the caption, you should explain the context and how the illustration should be interpreted, and draw attention to features you wish the reader to note, even if you have discussed these in some detail in the text.

- Don't cram in too much detail. When I ask students their view on the functions of tables, for example, they often reply that it is to record data such as experimental readings in a systematic way. This being so, a table might have to contain large amounts of data, perhaps extending over several pages. In my view such data should not go in the main text, but rather in an appendix. A table in the main text must be a complementary channel of communication, and illustrative rather than exhaustive—that's why they're called illustrations.
- Reserve the use of illustrations for things that are important. The reader will focus on them and assume that they are the most significant part of your work; use of illustrations for minor outcomes can skew the reader's understanding of your argument.
- Put in a table only when the patterning obtained by arranging things in rows and columns tells the reader something better than or different from a written description. If the data in your table seems to you to demonstrate some trend or correlation, you should consider displaying the trend by means of a graph, and banishing the figures to a table in an appendix.
- A diagram should be a net aid to understanding. If the work of explaining a diagram that illustrates, say, risk factors in diet is more work than simply explaining the risk factors, then the diagram is an unhelpful burden and should be discarded. (It can be helpful, though, to develop such diagrams for your own benefit, as they can clarify your understanding and help you focus in on what you are trying to say.) That said, never overlook the possibility that a pertinent diagram can greatly improve your presentation.

There are three kinds of figures: diagrams, graphs, and images such as photographs. This book is not the place to give detailed advice on preparation of such materials, but it is important that you be aware of some general principles.

Some authors like to describe aspects of their work as line diagrams made up of boxes, circles, arrows, labels, and so on. Such diagrams can be a powerful way of explaining relationships, but they are inevitably simplifications of complex situations, and may degenerate into caricatures. A diagram may be a useful way of illustrating the biology of food digestion, for example, but that doesn't mean that a diagram is the right way to show the connection between the sociopolitical factors that control foodstuff quality.

Graphs are used to demonstrate trends or correlations; so you need to think carefully about what you are trying to demonstrate. Usually you will be either confirming an established model or developing a new one, and you should have this in mind when plotting your graph. A common error is the introduction of an extraneous variable. In a striking example of this, the authors were trying to demonstrate that reducing the lead emissions into urban air from the combustion of petrol would reduce the lead concentration in the blood of children. The data available to them were figures taken in the US over the five-year period 1976–1980. The writers plotted both the quantity of lead in petrol sold in major cities and the lead content of the blood of children in those cities against time in years. The two curves followed each other almost exactly. The obvious (but wrong) conclusion that the unwary reader would draw is that lead in blood is proportional to lead from petrol, with the corollary that all that one had to worry about in a program to control lead in blood was reducing the lead emissions from burning petrol. If the authors had plotted one against the other, without worrying about the distraction of the years in which the various values were generated, they would have found that the correlation was much less pronounced, showing that, although lead from petrol was an important contributor to lead in blood, it was not the only one.

With the ready availability of rich tools for creating graphs, such as three-dimensional and coloured plots, the challenge for the author is to repeatedly ask the questions: Is this element necessary? And it is tasteful? No-one wants to read a thesis where every page is a blaze of iridescent colour like some sort of firework display. The more detail, the less likely that it will be correctly understood. My advice is that you should test your graphs (and other diagrams too of course) on someone, and listen to what they have to say. If they don't understand what is going on, simplification may be required.

Despite this availability of plotting tools, some people, mysteriously, still take their data and plot it by hand using software that is intended for creating pictures and diagrams. Do not do this; it will look ridiculous (or fraudulent) to every experienced reader.

Some theses in lab disciplines used to include photographs of equipment such as assemblies used for preparation of chemicals; maybe they still do. I am not persuaded that such photos are more useful than a diagram, and, in general, while I've seen a good number of cases where the photograph was included but seemed unnecessary, I can't recall a case where I thought more photos were required. Sometimes it is clear that photos are necessary (images of healthy and unhealthy cells, for example, or captured images of computer screens that demonstrate a software interface) but, for example, is it really appropriate to include photos of the teenagers interviewed about their eating habits? Maybe, but probably not; we all know that unhealthy eating can have consequences for appearance, and such photos can pass judgment on the individuals and suggest that the researcher has biases that may undermine the objectiveness of the research.

A picture speaks a thousand words, we're told. What words does a poor picture speak? One thing that really bugs me, and forgive me if I have said this already in some form, is the fact that some students appear to be comfortable with including truly appalling illustrations in their work. I am always astonished by the students who labour for days or longer over a fragment of text but are comfortable with a jumbled, clumsy picture that doesn't really illustrate anything. Unfamiliarity with tools is certainly part of this problem, as is the 'but I am not an artist' excuse. Perhaps they say to themselves that artwork is out of their expertise, and use this as a reason to quickly sketch something without even seeking advice.

The elementary tools for drawing figures and graphs with which most students are familiar, when used in an elementary way, are designed to be used by people whose use of computers is, well, elementary. They are used by children even before they can read and write. That is not to say that these tools can't be used better—but their default settings are certainly not intended for pictures that are to be included in research publications, and maybe it is a good point to remember that *every* paper ever published in a major journal over the last 350 years or so can still be found in a library somewhere. The vast majority of them are online at a permanent URL; your bad artwork can survive for a long time.

There are many specific things that I find jarring. This list is far from exhaustive:

- Lack of principles. Does a box represent an individual, or a collection, or an action? What is the difference between a black arrow and a coloured fat arrow? Do the colours have any significance? Why so many fonts and font sizes? Why are they so different from each other, and so much bigger or smaller than the regular text?
- Congestion. Lines that cross each other unnecessarily; arrows that end in space, or just inside the thing being pointed at; lines that might be pathways, but might be part of a boundary. Things under other things.
- Clip art. Are comical sketches (drawn by someone else, and often not very good) really what you want as the most visible feature in your thesis?
- Badly rendered photographs.
- Graphs with grid lines and unnecessary boundaries, poorly captioned legends, and too many lines. Missing data, out-of-proportion marks and line widths, poor use of space, and inappropriate sizing can all make a graph impossible to read, or worse, impossible to take seriously.

There are many good software tools for presentation, some of them free, in addition to those that are included in the common word-processing packages. Take the time to ask around and evaluate the options; you may be surprised at how much difference there is between them.

Many of the principles noted above for figures are just as applicable to tables. They should be self-contained, with detailed captions; they should not be amateurish. Choices need to be made about what grid lines to include or omit, how to align data in columns, what is important, and so on. Complex tables can present particular difficulties, when, for example, the data is multi-factorial but needs to be represented on a two-dimensional piece of paper.

I've found that a great way to locate examples of good tables is to leaf through a first-rate journal. These journals use professional typesetters to arrange tables, and have deep experience in which table elements are clarifying and which are unhelpful. This is again a case where the defaults in word-processing software are poor—indeed, I would say they are outright dreadful. Find good examples from your discipline, and imitate them.

A final thought on this topic is that many of the problems I see in figures and tables are, I suspect, due to the fact that authors don't anticipate how much work they are going to be. Given that the illustrations will be a conspicuous element in your thesis, expect them to require real effort to get right.

Summary of Chapter 8: Outcomes and Results

Research results:

- Data comes from sources and experiments described in earlier chapters. Only include data that is derived from a process that you have described, that is, the reader must understand where your data comes from. Describe the data fully.
- Have clear criteria for inclusion and exclusion of data and results. These should be independent of what the data shows, that is, it is not acceptable to only include data that confirms your hypotheses!
- Make sure you have used the right kind of analysis mechanism for your data. For example, tools or approaches for large data sets may be unsuitable for sparse or irregular data.
- Understand your variables.
- Build a clear argument from data to knowledge. As you build this argument, be aware that interpretation of the results may lead you back to the data collection process.

Presentation:

- Do not include raw, undigested data in the body of your thesis. Put it in an appendix, or better, back in your filing cabinet.
- Display your results in an informative, appropriate way, either through charts, tables, diagrams, or carefully constructed arguments. In doing so, make sure that the presentation makes it possible for the reader to see whether your hypotheses have been tested or your questions answered.
- Be open about shortcomings or limitations of your data or results.
- Figures should be reasonably self-contained.
- Use examples from elsewhere to guide your design of illustrations. Don't be content with word-processor defaults, which often look unprofessional, and use the right tool for the task.

Online resources:

- There are excellent web pages with examples of illustrations—though choosing the right query to find them can be a challenge; 'data visualization' worked well for me for graphs, for example.
- Wikipedia lists graphics software packages. Some of the best packages require that you write scripts in simple programming languages.

Chapter 9
The Discussion or Interpretation

Phil had reached the discussion chapter of his thesis, and we were talking about how he might shape it. The aim of his PhD project was to determine whether agricultural forestry could make a worthwhile contribution to the rehabilitation of degraded tropical uplands. He had spent a year on field research in Sri Lanka and, in the end, carried out two major research programs: a comparative study of existing land uses, and an economic analysis of a particular agroforestry system sponsored by a German aid agency. He had written chapters of his thesis describing the results of these studies, as well as the appropriate background chapters, but now found himself in trouble trying to pull it all together. Luckily, he had followed some earlier advice to keep his introduction and conclusion in alignment. I asked him whether he knew what the overall conclusions of his research project were. 'More or less', he replied. 'Enough to write them all down?' He hesitated, but 'yes' was the eventual answer.

When you have analyzed your results, you are much more proficient than when you started your project, but you now have to establish what can be concluded from these results. This is where you can advance from information to knowledge, where you might be able to establish new theories or new ways of looking at things. This is the task of the discussion chapter. Once Phil realized that he knew what he had concluded (more or less!), the task of writing the discussion that enabled him to get to these conclusions suddenly seemed less formidable.

Here I suggest a method for tackling the discussion chapter that makes it relatively easy to write. In many theses, there isn't an explicit discussion chapter, but instead, for example, the experimental chapters may have a discussion section. For convenience here I treat the discussion as a discrete chapter, while noting that it is often a component that might be included into the thesis in any of several ways.

The Task of the Chapter

Why does the discussion chapter worry students so much? The reason appears to be a variation on the problem I tackled in Chap. 4, that of the tension between the creative and the rational parts of our brains. In the discussion chapter the creative

D. Evans et al., *How to Write a Better Thesis*, DOI 10.1007/978-3-319-04286-2_9,
© Springer International Publishing Switzerland 2014

part of our brain is paramount, because we still have to compare the results of our own work with what we might have expected from existing theory to see what new ideas will emerge. Starting to write is, therefore, stepping into the void. Students often try to start their discussion in this way: they thrash around with a hodge-podge of undifferentiated thoughts in their heads, hoping that something will turn up. Yet we know that we must argue the discussion very tightly to convince the reader that the conclusions we draw at the end of the thesis are sustainable. This conflict must be resolved.

A related, and surprisingly common, problem is that some students are reluctant to state a definite view. Randall was uncomfortable writing down judgments on other people's work, although he was relaxed about making criticisms of the same work in our weekly meetings. 'What if they disagree?' he would ask. 'What if I offend them? They won't be happy if my work contradicts theirs.' This reluctance even extended to stating his own conclusions, because he did not want to be seen to assert that his work was better than other people's—yet most of us would see this as the whole point of doing research![1] Part of the problem was lack of confidence in the rational processes he had followed. (Another problem was a tendency to be excessively deferential to more senior researchers.) The first step to addressing this problem was to clearly identify what the conclusions were, and then to gain confidence that they were correct.

I advocated in Chap. 4 that you resolve the creative-logical tension by composing a rational structure for the thesis that will get you systematically from the aim to the conclusions. Once this structure is in place, you then start to flesh the argument out, giving the creative part of your brain free rein. Your writing might require that you modify the structure, or you might leave the structure intact and modify the argument. You have set up a fruitful dialectic. I also noted that you can use exactly the same technique when writing individual chapters. Each chapter must have an aim and conclusions, and you must structure the chapter in such a way as to get you logically from the aim to the conclusions. In most chapters it is not too difficult to do this, because you know what the conclusions are before you start to write. However, in this chapter research is still in progress, so you're not sure what the conclusions are. Therefore you can't design a structure that will enable you to reach them.

I've heard the discussion chapter explained by analogy with painting. Throughout your research, you have been close to the detail of the work (creating and touching up individual elements, standing right in front of the canvas), tightly focused on individual items. When you come to the discussion, you may be looking at the whole for the first time; it is as if you had returned to your painting after a break, and are now gazing at it from a few metres away, with an opportunity to assess and describe its strengths and faults as a complete work.

This is probably the part of your thesis where it is most important that you show your ability as a critical thinker. Examiners are particularly impressed by

---

[1] Randall was another example of a kind of research personality I mentioned earlier, with a reluctance to bring work to that final point where it could be refereed or examined; to complete something. In some students this is a striving for an unnecessary perfection or completeness—for example, a perception that because the work has opened up new questions then it is unfinished

candidates who are alert to shortcomings and limitations in their own work; indeed, why respect someone who shows no critical insight and seems to think that their accomplishments are without flaw? Step back, ask the questions of your work that you would ask of the work of other people, and use the answers to make your discussion penetrating and insightful.

Structuring the Discussion

How can you design a structure for the discussion that will enable you to get logically to your conclusions when you don't know what they are? Indeed, if you had *no idea* what the conclusions were, it would not be possible.

The resolution of the paradox is simple—we are asking the wrong question. When we assume that we don't know what the conclusions are, we are only partly right. The rational part of our brain is telling us that we don't know what the conclusions are because it knows that it is the function of the discussion to *find out* what they are. But the creative part of our brain has been working on this problem ever since the research project began. It has been trying out ideas and associations, sometimes accepting, sometimes rejecting, sometimes getting it right, sometimes wrong, but seldom informing the rational part of our brain what it has been doing, or where it has got to. Without this, research would not be possible. We have been doing research in our unconscious, creative minds all the time, and we have reached unconscious, creative conclusions.

The key to writing the discussion is for you to bring these unconscious conclusions to the conscious realm, and commit them to screen or paper. Your rational brain can then sort them out and do its best to make sense of them. You can then use them to design the structure of the chapter on the assumption that they *are* the conclusions. This is how to do it:

- Begin by brainstorming. Write down all the things that you know now that you didn't know when you started the research; a single sentence for each item. These can be big ideas, little ideas, snippets of knowledge, insights, answers to questions, whatever. Don't worry about whether you are responding to the aim you set yourself in your introductory chapter. That would be a rational approach, whereas you are engaged in a process of dredging up unconscious conclusions. Consider asking your supervisor or a colleague who is familiar with your work to sit down with you while you are listing these conclusions. The presence of another person, chipping in and asking questions, may help you to uncover your hidden thoughts. You should end up with a totally undifferentiated list of maybe 20 or 30 'conclusions'.
- Sort these into groups of associated ideas (now using your rational brain). You will probably end up with three or four groups. If you have more than four groups, you may have included conclusions that emerged earlier as conclusions to your background chapters, but which have not interacted with your own work.

Perhaps these were important at the time, but they are not conclusions from your whole research. You may test this by asking, 'Does this conclusion respond to the aim I stated in my introductory chapter?' Reject those that don't, but first check that you did give them clearly as conclusions in the earlier chapters where they belonged. If you still have more than four groups of conclusions, try coalescing them.

- Give a heading to each group. These headings will form the section headings in your discussion chapter. The function of each section is to argue for the conclusions that you will be drawing later. Examine these headings to see which order they should go in. (You will find that some of the groups of conclusions don't make too much sense unless you have already dealt with others.)
- Each section will contain several points, as identified by the separate conclusions that you have already listed for that section. These could form sub-headings within the section. Sort these sub-headings into a logical order, reject ones that are obviously irrelevant, add others that you now see you missed by your earlier haphazard identification process, and coalesce points under one heading if this makes sense (you should not have more than three sub-headings within a section).

You will now have a tentative structure for the discussion chapter and may give your creative brain permission to write the text, using this structure as a framework. When you start to write, you will not be stepping out into the void.

This balancing of the rational and creative parts of our brains by writing creatively to a rational structure will work only if you treat it in this way. There will be an ongoing tug-of-war. Often your creative mind will take you away from the rational structure. When this happens, don't assume that the creative mind is always right. Similarly, don't assume that the rational mind is always right. But you cannot leave it unresolved: you must bring either the structure or the wayward text into line. This problem will be particularly acute in this chapter, because the rational structure you are using is tentative, being itself based on conclusions garnered from the creative mind. However, our experience is that at this stage of the research the creative mind has already done marvellous things, and usually you won't have to change the structure much, even though you may modify some of the individual conclusions.

The process I have described probably seems messy, with much experimentation and correction to do. Messy it is, but this is the chapter where research is still going on; it is the main one in which the act of writing might cause you to find out more, where knowledge might become wisdom. Nevertheless, you may find it relatively easy to write the chapter because this chapter is the one that all the others were leading to.

Remembering Your Aim and Scope

A function of the discussion section is to respond to the aim you set in the introductory chapter. Before they start a detailed reading of your thesis, most examiners will flip from your introduction to your conclusions to see how your concluding ideas line up with your original ones. They have been asked to do this in the suggested criteria for examination: Are the conclusions and implications appropriately developed and clearly linked to the nature and content of the research framework and findings?

Examiners criticized me (Gruba) on this very point. I had not made links from throughout my discussion section to the overall conceptual framework that I had introduced in my review of current theory. When I got their comments back, I had to return to the world of abstract ideas after a long sojourn in the analysis of my own data set. What I had to do was to pull back from the results in order to reflect on their underlying concepts. I found that I had to re-read my background chapters, refresh my understanding of the tone and style of the discourse and write in such a way that I was responding to the major points I myself had identified. This is not an easy task, but you must place your own study within the larger research framework.

Quite often students find that their work has gone much further than they originally dared to hope, and their original introduction, including their stated aim, has not caught up with this.

Writing with Authority

By now you have earned the right to comment on the field, and you can (and must) do so with authority. How can you demonstrate this authority? In this chapter you need to address three areas with a critical eye: current theory, current practice and the conduct of your own study.

First, you should make sure that you place your thesis within the context of the field you are working in. In addition to making links from the research framework to your own study, you now have to suggest ways to expand that theoretical point of view. To start, I suggest that you question or illuminate the accepted definition of potentially controversial key concepts and phrases.

The expansion, or possible contraction, of existing categorizations of key factors in your field is another area you should consider. For example, the results of a project on the way refugees access welfare services in Australia might suggest that we need to go beyond financial and medical problems to include family problems. Or the results of a study aimed at developing plans for recycling might indicate that city planners should consider personal and social identity, which would call for an examination of how this might be incorporated. One of your major contributions to the field will be the development and discussion of such factors.

Because you have earlier developed an awareness of the limitations of current practices both through your review of earlier studies and your own application of them, you are now in the position to suggest ways to improve them. What would you have done differently, and why? Here you can act as a guide for further researchers. Tell the readers what worked well, and what did not.

In many theses you will find a section entitled 'Limitations of the Study'. Whether you put this in a separate section or discuss it where appropriate as you go along, you must deal with it. This section need not be set in an apologetic tone; rather, it should acknowledge areas that you yourself thought were weak and deal with them in a straightforward way. It is where you show your ability as a critical thinker.

For example, in her study of 'Indigenous Development Projects', Perla used as a case study a nutritional project in the Philippines designed and managed by local people. Among other things she compared the views of the project held by two groups of villagers in a case-study village, participants in the project and non-participants. The difficulty of reaching the non-participants and communicating with them in the local dialect limited the number of respondents in each group to 15. This number is generally held by statisticians to be too low to get statistically significant differences between the average values of parameters for such comparative studies. She understood this, but her limited resources made it impossible for her to interview more respondents. She was careful to say what she should have done in an ideal research world, and to acknowledge the limitations imposed on interpretation of her results. Examiners will understand that you have worked as an individual researcher with limited time and resources and, in most cases, will not reprimand you for 'not doing enough'. However, they will be critical if you show that you are unaware of key problems in the management and execution of your research.

Another example is the 'case-study' problem already mentioned in Chap. 7. In case-study work you are examining what you have chosen as a typical example (or examples) of an activity or a place and examining it in some detail. Because of the detailed examination you are often able to derive some strong conclusions. But how well will these conclusions apply to other similar activities or places? This may be difficult to establish, but you must attempt it. (Perla spent 18 pages on this problem.) If you omit discussion of the problem of generalization, you are saying to the examiner that you are either unaware of what a case-study approach is attempting to do, or that you are aware but are not confident of trying to make such an assessment.

Overall, it is in this chapter, in many theses, that you establish yourself as expert. You have proposed a hypothesis, established a method for evaluating it, undertaken experiments or measurements or a review, presented your results, and analyzed them against clear criteria—in other words, made a contribution based on your knowledge and data. Your presentation of the arguments is where you persuade the sceptical examiner to your point of view, so it is essential that you be cogent and clear. With a good presentation, you will truly have demonstrated authority in your field.

Summary of Chapter 9: The Discussion or Interpretation

Structuring your discussion:

- The task of the discussion chapter is to enable you to reach your conclusions. Drawing up a tentative list of conclusions will help you identify an appropriate structure.
- Begin by writing down all the things you know now that you didn't know when you started the project. Rearranging this list will give you the titles of the main sections of your discussion.

Checking the thesis structure:

- Before you start writing material in each of these sections, check your thesis structure by stringing together introductions and conclusions for all the chapters.
- Check that the tentatively structured thesis responds to the aim and scope you set yourself in your introductory chapter.

Write with authority:

- Make sure that your exposition of new theory or ideas places your thesis within the context of the field you are working in. This will require that you not only draw on your own results, but that you view these against existing thinking as expounded in your background chapters.
- Acknowledge any limitations on your findings. Theoretical results may need validation before their suitability in practice is known, for example. Shortcomings or uncertainties should also be acknowledged.
- If the thesis involves a case study, check that you have dealt with the problem of generalizability, or issues of transference, for your findings to similar situations.

Chapter 10
The Conclusion

You stated the aim of the research project in your first chapter. These conclusions must indicate how you fulfilled that aim, and must arise inescapably from the argument in the discussion chapter. Researchers often state conclusions that they have failed to argue for. They had become convinced of them in the course of their research but, because they did not follow a process such as the one I described in the previous chapter for structuring the discussion, they had omitted to back them up in their writing.

It is essential to forge the links between the 'Introduction' and the 'Conclusions', and between the 'Discussion' and the 'Conclusions'. As these are the conclusions to the 'Discussion', it follows that the discussion chapter does not need its own separate conclusions. (For this kind of reason, I find it clearer to have end-of-chapter summaries rather than conclusions; or it may be that you have structured each chapter so that it ends with a section headed 'Discussion', rather than 'Conclusions' or 'Summary', and have omitted the 'Discussion' chapter entirely. For convenience here, I've assumed that you have a separate chapter devoted to discussion, but I don't think it is necessarily superior to having a discussion in each chapter, especially if your thesis is a series of linked contributions.) You could roll the discussion and conclusions into one final chapter, giving it the title 'Discussion and Conclusions', as is sometimes done in papers for learned journals, but I think a separate chapter of conclusions is much preferable.

If you followed the suggestion I made in the last chapter, you will have a set of conclusions that emerged out of each section of your discussion, rather than the ones that you dredged out of your unconscious mind when you started the procedure. You can now write these down as the conclusions to your research, knowing that you have argued rigorously for all of them, and that you have got them in perspective through your argument. Also, if you put them down in the order in which they emerged in the discussion, they will be in a logical order, because you arranged the discussion in a logical order.

In some disciplines it is customary to use the conclusions chapter to briefly examine the implications of your findings, such as their likely impact on future work, on other research areas, or on practice of a profession. In some fields, examiners expect that your thesis conclude with an agenda for further research.

D. Evans et al., *How to Write a Better Thesis*, DOI 10.1007/978-3-319-04286-2_10,
© Springer International Publishing Switzerland 2014

One way that I get my students to think about an agenda for further research is to have them picture their responses to questions at a post-doctoral job interview. 'Now that you have completed a thesis', an imagined professor asks, 'what do you expect to work on in the coming 5 years that will contribute to this department?' With that question, or something similar, we then begin to brainstorm how the work may be extended. Will you focus on its theoretical implications, and what future researchers may need to think about as they begin to work on their own conceptual frameworks? Does your work have 'practical' application in that it illuminates the processes of building faster, more efficient processes and machines? How does it relate to the development of professionals in the discipline: Should they be told, for example, to train more on food safety in an era of increased litigation? For many students, working on an agenda helps them to see the implications of their work in a wider context. I have seen it be enormously validating for some students.

Note, however, that it is fine for such a discussion to be speculative, but it shouldn't be fanciful. It may also be appropriate to sketch future research directions that your conclusions imply, or to suggest additional work that your investigation left incomplete. When it is poorly done, such a discussion can appear absurd; when it is done well, such a discussion can powerfully communicate to the examiner your understanding of the importance and limits of your work.

You should now have a deep sense of satisfaction about the whole thesis! Any residual doubts may indicate that something is wrong earlier in the thesis, and you should try to find out what it is. Some diagnostics are in the following rules.

- I have already hinted at the first rule. If the discussion chapter is where you draw together everything you have done in your whole research project (not just your own experiments or surveys, but also your reviews and analyses of the work of others), then *you should draw your conclusions solely from the discussion chapter*. If you find yourself wishing to include conclusions that you have not worked over in the discussion, you have either omitted something important from the discussion or, more likely, you are still hankering after more than one aim.

- *There should be only minimal discussion in the conclusions chapter*. If you find yourself wanting to engage in further discussion, and are even still quoting from the literature, you should have incorporated this material in your discussion chapter. The conclusions can be a good place to tightly link together the themes that have emerged in your thesis, but a detailed analysis should take place elsewhere.

- *The conclusions should respond to the aim stated in the first chapter*. If you take your problem statement and then the aim from your 'Introduction', and follow these with your 'Conclusions', the result should be a mini-document that reads logically. When looking at the first draft of a thesis from one of my own students, or examining a thesis from some other student I always put it to this test. It often reveals that the writer omitted to state the aim, and it is only when one reads the conclusions at the end that one can start to deduce what the unstated aim must have been.

- *Summaries are not conclusions.* I drew this distinction in Chap. 2 when talking about conclusions to individual chapters. It was important there; it is even more important here. Repeating what I said earlier: summaries are a brief account of what you found out; conclusions are a statement of the significance of what you found out—what you concluded from it. If you are merely summarizing the argument developed in your discussion chapter, you will feel quite unhappy with your conclusions. There will be no sense of closure. Also, you will almost certainly have failed to respond to the aim of the whole project. (Sometimes this happens when the aim is too modest, or even woolly. For example, when researchers say that their aim is to investigate the properties of a system, they may end up with a list of properties, a summary. This is hardly research.)
- *Conclusions should be crisp and concise.* The conclusions chapter may be only two or three pages long—which helps to give the sense of closure mentioned above.

Summary of Chapter 10: The Conclusion

Connecting discussion and conclusions:

- Your conclusions are what your discussion chapter has been arguing for.
- You could write the conclusions to your whole study as the last section of a chapter called 'Discussion and Conclusions', but it is usually preferable to have a separate 'Conclusions' chapter.
- If they form a separate chapter there should be no conclusions to the discussion chapter, and you should inform the reader of this in the discussion.

Rules about conclusions:

- You should draw your conclusions solely from the discussion chapter.
- There should be little further discussion in the conclusions chapter.
- The conclusions should respond to the aim stated in the first chapter.
- Summaries are not conclusions.
- Conclusions should be crisp and concise.
- The conclusions can be used to briefly explore the implications of your findings.

Chapter 11
Before You Submit

You have just typed the last full stop of your conclusions. Finished at last? Wrong—you still have many weeks of work to do. You have two major tasks ahead: you must revise your first draft in response to the criticisms of supervisors and friends and, when you have done that, you must check the details of the whole work.

What you have actually finished is your first draft: a collection of chapters written according to the structure you devised. Now you will need to really focus on 'structural editing'. At this point, it may appear to you (and your supervisors) that each chapter is coherent, but you now have to consider whether the whole thesis hangs together. You will also need to check whether your argument really gets you from the aim to the conclusions; whether your aim itself has drifted during the course of the research; whether there is extraneous material that you should transfer to appendices; and whether important insights have emerged to which you gave little or no prominence in your original structure, but which are now demanding more attention. When you have put all these things right, you will have completed the structural editing.

As you work through the second draft, you will also need to work on editing details. There are many aspects to check: spelling, punctuation, captions to figures and tables, references, and consistency in everything including nomenclature, format, reference style, and writing style. Although these things are not always intellectually demanding you have to do them properly, and they take time. And they *are* important. What may seem to you to be minor errors in your text can send the reader strong subconscious messages that, basically, you don't care and that the work is slipshod. This is not a message that you want the reader to receive.

From First to Second Draft

Some graduate students, when they have typed that last full stop in their conclusion chapter, print out a clean copy of the entire thesis and give the whole thing to their supervisor to read. Although this gives students a strong sense of completion,

D. Evans et al., *How to Write a Better Thesis*, DOI 10.1007/978-3-319-04286-2_11,
© Springer International Publishing Switzerland 2014

and of self-congratulation for all of the hard work they have put in, there are reasons why this is not a good idea. First, the supervisor should have been giving feedback chapter by chapter, and may already have expressed complete satisfaction with some chapters while asking for extensive revision to others; if a supervisor thinks a chapter is done, there is no need to ask her or him to read it again. Second, quite possibly your supervisor should only see your complete thesis after any other reviewing is complete—in particular, after *you* have reviewed it. If you can see that further revision is necessary, why waste your supervisor's time doing work that you can do yourself? I recommend this not only because it is a unique and necessary experience, but also because the comments that you get back from your supervisor from a document that is in good shape will be more useful than the comments from one that is still full of problems.

Also, as noted earlier, your supervisor has many commitments, while you only have this one. Handing over the thesis chapter by chapter means that you can continue to work while you wait for the feedback; a supervisor who is given a complete thesis may not return the manuscript for months.

What I do with the first draft is parallel to what I expect the examiner of a thesis would do, or what I would do if I were refereeing a paper submitted to a conference or a learned journal. The only difference is that, because I am your supervisor, I am now fairly familiar with the drift of your argument and with the approach you have taken, and I have to guard against reading things into the draft that you have not clearly explained. When you are reading your own work, this is even more of a problem. For that reason, you should put it aside for a few days before you read it as a whole.

When making a detailed review of this kind, I prefer to work on a printed copy rather than at my computer. Despite over 25 years' experience of writing documents digitally, I find that a complex document cannot be read thoroughly in electronic form. As word-processing software improves, the advantages of hardcopy are reduced, but it is still the case that it is easier to page through a document and annotate it on a printout than on a screen, and the brightness and relatively low resolution of screens make them much more tiring for extended reading sessions.[1]

On the topic of supervisor feedback, by this stage you should have formed your own judgments about how reliable it is. Some supervisors are extremely careful and give specific advice on what to fix and how to do it. Others tend to give generic advice[2] that can lead an incautious student to make a mass of unnecessary changes. A colleague of mine was tormented by his supervisor's habit of asking for a change on one draft, then, on the next, asking for it to be changed back. The problem was that the supervisor sometimes didn't take the time to read the work properly, and thus didn't appreciate why things had been presented in a certain way. The lesson here, as in much of this book, is that you should be sceptical and think for yourself

---

[1] However, my revision of this book was not printed until I had a final, 'clean' manuscript.

[2] The dreaded 'rewrite!' was a comment that was often written on my thesis. I cursed my supervisor every time I saw it, because he used it for everything from minor errors of punctuation to major garbling of whole passages, with no hint of what the comment referred to.

about everything, not blindly follow what may be wrong information or bad advice; especially when you *know* it is wrong.

Sometimes, specific advice is not appropriate; if I see some sentence or argument I don't understand, but where I suspect the student hasn't really thought about what they are trying to say, I may simply annotate it with a comment such as, 'Are you sure you know what this means?', and my students (are expected to) understand that fixing it means that, first of all, they need to try and analyze, then rectify, the problem for themselves—but also that I expect them to check with me before doing anything drastic. If the problem is subtle or complex, I'll also include an explanation, because while I do want my students to develop their critical thinking skills, I don't want them to waste their time.

My student Kari wasn't good at handling feedback from me. She was not an experienced writer and, although she produced text quickly, it was often full of mistakes. Worse, it tended to be disorganized with bundles of unrelated thoughts gathered into the same paragraph, or the same topic discussed in multiple places, or even whole paragraphs amounting to hundreds of words repeated in different sections. Indeed, this was something like her style in conversation! She had made useful discoveries and, when pressed, could explain in an entirely coherent way exactly how the results and hypotheses related to each other, but in her writing (and speech) she often seemed to be gathering her thoughts and reaching conclusions as she went along.

By itself, I did not see this way of generating text as a problem—her approach certainly helped her to make interesting connections and guesses. What was a problem was her lack of understanding that the resulting 'brain dump' was unreadable. In one particularly trying instance, I spent several long evenings marking up one of her chapters in a great deal of detail, in the hope of explaining to her how to reduce her rambling but informative text to something more punchy and concise. The feedback was in terms of grammar, word choices, organization, flow of ideas, and comments on missing or unnecessary text, which we reviewed together in a meeting. But her ego had been hurt, and after our meeting her response was to throw away the draft, including all my comments, and start again! I hadn't made a photocopy (another lesson learned) and between us a great deal of work was lost. The new version was not much better than the original, and, though it was hard to be sure, I felt that some of the insights were forgotten. I later found out that she had decided that my extensive comments—there was a *lot* of ink on her draft—were a way of telling her that the manuscript was rubbish. In other words, she overreacted. On a smaller scale, I suspect that some degree of overreaction to supervisor feedback is common.

My student Louis made a more elementary mistake. He would make changes based on my written comments as precisely as he could, even when he couldn't decipher them or, on reflection, they didn't make sense. It was as if he was afraid of offending me by deviating from my observations; he had not understood that a supervisor's comments are not instructions, but guidance.

Structural Editing

First, I look at the overall structure. There should be a table of contents that corresponds with the chapter titles and main section headings in the text.

The table of contents should tell me straight away whether there are any major logic problems. If it is not informative enough, I go to the beginning of each chapter and read the introductions in order. This will probably help, but it may reveal that the introductions themselves are inadequate. (I will already have done some reviewing of this kind while the thesis was in early stages of preparation).

Finally, I read the introductory chapter *as if I were a reader seeing it for the first time.* I ask myself: Is this telling me (the uninformed reader) why the work is being done? Is it clear what the aim of the work is? Is there an adequate sketch of how the writer intends to achieve this aim? Is the scope of the thesis clearly delineated? Again, if any of these points are inadequate, I note the problems in the margin. Then I go straight to the conclusions, and ask myself whether they respond to the stated aim. If they don't, I note the disparity.

The Main Text

Next, I read the whole draft from beginning to end, noting spelling, grammar and typographical errors as I go, and also noting things such as obscurities, pullulating patches of purple prose, and places where the argument seems to have logic gaps. At the end of each chapter I write a few lines about how the chapter shaped up in the context of everything that preceded it. The conclusions or summary of the chapter are particularly important here. One of my most common comments on them is that the author is still writing a list of the chapter contents, rather than giving me, the reader, a sense of how the chapter is advancing my comprehension of the argument in the whole document.

A related check at this stage is to see whether the text has *flow* and *motivation*. When I am reading, say, an explanation of cellular metabolism, I want to know why it is there—how does it relate to the overall topic of teenage dietary choice? When I am reading a description of experimental apparatus, I want to know why that apparatus was required, and what kinds of tests it is going to be used for. The old adage is 'tell the reader what you are going to say; then say it; then tell the reader that you have said it'. Sometimes this is described as 'make sure that everything has a head and a tail'. At a coarse level, this is how a chapter is organized—the introduction and summary provide motivation and context for the chapter's contents. But the same principle should be applied at a more fine-grained level, so that the reader is never left wondering, for example, how a particular section or even paragraph could possibly be relevant to the rest of the thesis. While head-and-tailing can easily be overdone, it is a critical tool for helping to ensure that the thesis has narrative flow.

By the time I have reached the end of the thesis, a sense of the integrity (or lack of it) of the whole document has usually built up. If there is a problem, it may be obvious. If it is not obvious, I repeat the first step—the examination of structure—but now with knowledge of how the whole argument has developed; or has failed to develop. There may be major gaps in the argument; there may be material present that is not part of the argument and that should be relegated to appendices; there may be repetitions that should be eliminated or consolidated; there may be material that would have been better located elsewhere in the document; there may be conclusions emerging strongly at the end that the author should have emphasized more, or had failed to argue for in the discussion; and so on. Before handing it back to the author, I write a few pages on these larger problems.

Thus the author now has two sets of comments: detailed comments in the text on points of grammar and expression; and general comments about the structure of the argument. We discuss the latter, and the student gets to work on the second draft. As the student produces revisions of various parts aimed at solving particular problems, we discuss them. I usually find that a complete re-reading of the second draft will not be necessary until after the second, more detailed, part of the finishing process that I am about to describe.

Dotting the 'i's and Crossing the 't's

Although the second draft is now essentially complete, you still have some weeks of detailed, rather tedious work to do. Don't skip it—tedious or not, it is essential. The items that you need to check are listed below in the form of a check-list. You may even want to photocopy this checklist and tick the boxes when you have completed each task. If you have used your word-processing software to its fullest many of the jobs will already have been done.

Your thesis may include text that you already regard as 'finished', in particular material drawn from papers that were written during your candidature. Some students seem to think that such text doesn't need checking, but you would be surprised how much change can be required due to the need to integrate the paper into a complex thesis. Make sure that all text is checked to the same level of detail.

Preliminary Pages

The first few pages, before the start of Chap. 1, are preliminary pages that set the context of the thesis, and help readers to find their way into it. They will include some or all of the following, generally in the order given below:

Title Page

- Check that it contains title, author, place, month and year, and the degree for which the thesis is submitted, and any university-specific requirements.
- Check that the title of the thesis accords with your aim. If you decide to be clever, two risks may arise here: (a) You may be tempted to use an eye-catching title that could disorient the examiners. This might make you feel pleased with yourself, but it is better to make sure that your thesis passes! (b) The work is difficult to locate for future researchers because the title contains obscure, or perhaps very common, terms that result in unproductive searches for a reference.
- Check if you are officially still using the title that you nominated to the university at the start of your project. Almost certainly the thrust of the project will have changed over the course of your candidacy, and you should change your title accordingly.

Abstract

- Needs to be present, and should contain summaries of the three main components of the project, as outlined in Chap. 2, an individual paragraph to each: (a) why you did the work and what you were trying to achieve; (b) what methods you used and what results you obtained; and (c) what you concluded from it.

Table of Contents

- Have you listed all chapter headings and headings of main sections within chapters? (Many authors also list sub-section headings. I suggest that you don't; they clutter up the table of contents and rob it of the power to demonstrate the structure of your thesis).
- Have you listed all endmatter (References, Appendices, and so on)?
- Check the styles for table-of-contents entries, and change them if necessary to make a neater, more informative Table of Contents. It's a good idea to look over several completed theses to see how other students have done this. Then you can select for yourself what works well and avoid what works badly.
- Optionally, the Table of Contents includes lists of figures and tables; these are not essential, but are valuable in a thesis with a lot of technical material.

Preface and Acknowledgments

- A preface should give any information about the preparation of the thesis that you feel to be necessary, for example how you came to embark on the project. Prefaces are seldom necessary for theses. If you do have one, any acknowledgments should appear as part of it.

- Acknowledgments recognize help received in the execution of the research and in the preparation of the report or thesis.
- If you are fortunate to have received financial assistance, don't forget to acknowledge the organization that granted you a scholarship or other funding.

Declaration

- Most universities require certification that the work in the thesis is your original work, and has not been used for the award of any other degree.
- If you have published work from your thesis in journals before making a final submission, you must list complete references to such articles. You need to do this to avoid possible accusations of 'self-plagiarism' or submitting work that is not entirely original, and you will need to clearly identify the extent to which the papers are your own work as distinct from that of your co-authors.

The Main Text

- If you have been following the methods I advocated in the preceding chapters, everything appearing in this checklist should already have been done. But do check. If you have just picked this book up and have not been following my suggestions, I strongly urge you to use this checklist. If you find any of the suggestions puzzling, go back and read the chapter concerned.

Aim and Scope

- Can the aim be located in the table of contents?
- Is the reason for doing the work outlined?
- Does the aim follow clearly from this problem statement or rationale?
- Are constraints stated that limit the scope of the investigation?
- Is the aim followed by a brief outline of the way you intend to go about achieving it? (This refers not only to the experiments, surveys or investigations that you will design yourself but to the whole of the project, including reviews of theory and so on).
- Do the conclusions you draw in the last chapter relate clearly to your aim?

Background

- Do the introductions to chapters and sections clearly state their purpose?
- Is there any material in the background chapters that does not contribute directly to the later development of the report or thesis? (If there is such material, it should be relegated to appendices, or omitted altogether).

- Do the background chapters justify the formulation of the hypotheses or research questions?
- If you are using a case-study approach, does the reason for selecting the case study, and a description of it, appear among the background chapters? (It should not, as it is part of your research method, and such material should not be described until you have selected your method).

Design of Your Own Work

- Do your hypotheses or research questions spring logically from your reviews of theory or practice, or from your preliminary surveys or experiments?
- Do you discuss the possible methods for enabling you to test your hypotheses or answer your questions?
- Do you explicitly select a particular method or methods, and justify your selection through your review of possible methods?
- Do you explicitly design experiments or other research programs to implement the selected method or methods?
- Are tests for your hypotheses or ways of investigating your questions unequivocally built into your research programs?
- If you have decided on a case-study approach, have you justified this decision adequately?
- Have you justified the selection of your case-study activity or area in terms of its representativeness or typicality or other appropriate criteria?
- Unless offset by a colon and designated as such, does the name of the case study appear in the title or aim of your thesis? (It should *not*. If it does, you still have not sorted out the difference between a study of something in its own right, and the use of a case study to investigate something else).

Results

- Are the results of your experiments or surveys or other own work clearly presented and explained?
- Are displays, such as graphs, tables and figures uniform in style and numbered?
- Are the major trends or findings outlined? (You should not be discussing the implications of them while you are reporting them. For a short paper this might be appropriate, but for a thesis you should keep them separate).

Discussion

- Do you discuss your own findings in terms of their implications for one of the four areas of possible contribution, particularly with respect to modifying or extending existing theory or practice?

- Does the discussion permit you to reach all of your conclusions?

Conclusions

- Are all your conclusions justified by the preceding discussion?
- Are you forming new points for discussions while drawing your conclusions? (You should not be).
- Do your conclusions respond to your aim, as set out in your first chapter?
- Are your conclusions merely summaries of findings, or do they draw out the implications of your own work in terms of improved theory or practice? (They should).

Format

You will have to satisfy yourself that the format you have used helps readers to find their way through the thesis and, in particular, that it is consistent. Most books on writing theses give a chapter or more to this, with strict rules about the numbers of spaces before headings, the method for emphasis of major headings, the use of numbering systems, the spaces between paragraphs, and so on. Most such properties are managed by effective use of templates in a word processor, as discussed earlier.

Figures and Tables

Check all figures and tables. All will have a *caption* that should consist of several parts: a title (which will appear in the lists in the preliminary pages); explanatory material that draws attention to or explains certain features of the figure or table; and a citation giving the source of the material. You may lump all the figures, including graphs, diagrams, plates, photographs, and maps together in one list and the tables in another, although in the past it has been customary to make a separate list of photographic plates (a practice that predates the use of high-quality, computer-generated copies of photographs).

Any Figure or Table

- Does it add an extra dimension to your ability to give a piece of information, demonstrate a trend or communicate an idea?
- Is it simple or cluttered? Do the important points that you are trying to make emerge clearly?
- Does it, together with its caption, make sense by itself, or does the reader have to read the text to make sense of it? (One should not have to).

- Do you draw attention to important points in the caption?
- Is there a reference to the figure in the text *before* the figure itself?
- Have you acknowledged the source or the information on which it is based?
- Have you identified examples of 'good' figures in other people's work, and applied the lessons to your own work?

Graphs (or Charts)

- Does it have both axes clearly labelled?
- Is the text legible?
- Are lines and other features appropriately labelled?
- Have you sought out and followed guidelines on design and preparation of visual materials?

Tables

- Have you arranged it in some way that makes it more than a collection of data? Would the reader see patterns or trends? (There is no justification for having tables otherwise).
- Is it vertically and horizontally consistent?
- Are there unnecessary lines?
- Are all rows and columns labelled?
- Have you considered relegating the data contained in it to an appendix, and plotting the main trends as a graph?

Notes and References

If you have used the numbered notes system of references, and you have used your word processor to automatically number or renumber notes, you should not have to check that the note numbers correspond to the reference numbers in the text—this should have been done for you. It will enable you to collect your notes at the foot of each page or at the end of each chapter or at the end of the main body of text (but collected separately for each chapter), before your 'References'. Give your list a heading 'Notes', as a section-style heading if at the end of each chapter, or a chapter-style heading if at the end of the text.

However, there is still some checking to do:

- Re-read the notes to make sure that you have not deleted any accidentally, and that you wish to keep the ones you have.
- Check whether you need to revise any of them in the light of revisions to the text.

Whichever system you use, you should include a full list of sources such as papers in journals or chapters of books that have been cited. The list should be in alphabetical order of authors' surnames, and should contain sufficient detail to enable the reader to find the material in a library. You should check your list for three things:

- Is your reference list in alphabetical order?
- Do the entries conform to an established style?[3]
- Do all the references cited in the text appear in the list?

You should head this list simply as 'References' in the style of a chapter heading.[4] This list is often placed after the chapters but before any appendices, presumably on the grounds that the appendices really are something tacked on to the end. As the appendices themselves may have references, there is a case for reversing this order. If you leave the references in the customary place, you should devise some logical method for overcoming this problem, perhaps by having a short list of relevant references at the end of each appendix. Check to see that you have dealt with this problem adequately.

Appendices

You may have ended up with a rather mixed bag of appendices after completing your first draft. Some of them will have been written for very good and valid reasons to support material in the text. Others may be leftovers from earlier thinking, and because you were rather attached to them you were loath to throw them out.

- Check your appendices against these rules, and throw out any that are no longer justifiable.
- Check the presentation of each appendix that you decide to keep, as follows: (a) Does it start on a new page? (b) Does it have a title that indicates what it is all about? (Just calling it 'Appendix 3' is not good enough). (c) Is the style used for the title the same as that used for chapter headings?
- Is there a preamble that explains briefly what its function is and what it is all about?
- Does the preamble refer to part of the main text? If it doesn't, find the part of the text that it supports and make reference to it. If you can't find it, or if the connection is very weak, throw out the appendix altogether.

---

[3] The order in which the various components are given, and the styles used to distinguish book titles, journal names, and so on, varies from discipline to discipline. You should find the method used in your discipline and stick to it; be very consistent. Departments will often have a preferred method.

[4] Some people prefer to call it *Bibliography*. A list of references contains only material that is specifically referred to in your thesis, whereas a bibliography may contain other material of interest, but not specifically referred to. For a thesis, *References* is preferable.

Glossary

If you have a glossary, it is customarily placed at the end, after all the appendices. In other theses, it is placed near the lists of tables and figures, which may be part of the table of contents.

And don't forget: Your thesis needs to consist of your work, not other people's. If you have text that is drawn from papers that are co-authored with others, make sure that they understand that it is being used in your thesis. Only include figures and tables if you have permission to do so; if you are not the author of a figure, you must ask the authors and publisher if you can use it. If you have access to a tool for checking for plagiarism, consider using it. If you are including material from papers you have written, it should be material you were responsible for; if one of your co-authors wrote a section without significant input from you, then it should not become part of your thesis.

Summary of Chapter 11: Before You Submit

When your supervisor has 'signed off' on every chapter of your thesis you have only finished the first draft of your thesis. You still have two major tasks ahead of you: checking the structure and checking the detail.

From first draft to second draft:

- Check the structure of the thesis as a whole.
- Read it through in detail yourself. Check the logic flow. Look for gaps in the logic, repetitions, things in the wrong order. Fix these up to the best of your ability.
- Then (and only then) ask your supervisor to do the same. If possible, find a friend whose opinions you can rely on but who is unfamiliar with your topic to do the same. Fix up the problems they identify.

Checking the details:

- You now must check to ensure that you have done everything properly. A check-list is given in this chapter; you can find other such checklists online. Depending on how systematic you have been earlier, this task may take several weeks. Allow time for it in your thesis completion schedule.
- Be professional. Do not use material that is not your own without proper citation, and be aware of ethical concerns that lie at the core of academic research.

Chapter 12
Beyond the Thesis

Students enrol in a research degree to develop as researchers, make discoveries, and, ultimately, write a thesis. More broadly, a fundamental goal underlying research study is that of transformation, from knowledge consumption to knowledge production, from dependence to independence, from student to colleague. You would not be doing a graduate degree unless you thought it was going to lead you into the world of professional investigation, or of research and scholarship.

This is not to say that most students do research because they are ambitious in terms of career. That may be true for some—and it is a perfectly sensible motivation—but, for many, the degree is a consequence of a desire to do research, to work with people they admire or identify with, or to be in an exciting workplace. Nonetheless, at some point, probably quite early in your studies, you would have asked yourself: 'When this is finished, what next?' In my view, this is almost the same as another question, which has an explicit notion of transformation: 'Who do I want to become?'

Such a transformation may be primarily driven by the work that is captured in the thesis, but also encompasses development in a range of other areas. By the end of their formal study, a typical strong PhD student will have published a paper, or several papers in journals or at conferences; presented seminars, possibly in front of intimidating senior audiences; taught undergraduate classes; reviewed papers, either independently or under the guidance of a supervisor; visited some other academic institution; and, perhaps, mentored some undergraduates and junior research students.

By submission time, you may well be planning further steps in all of these areas, and perhaps even contemplating writing a book or obtaining your own funding. On the other hand, you may be looking forward to work with an employer whose agenda does not include scholarly work, and your PhD candidature might be your one chance to get your research published. In both cases, then, you need to be planning to publish as you study.

D. Evans et al., *How to Write a Better Thesis,* DOI 10.1007/978-3-319-04286-2_12,
© Springer International Publishing Switzerland 2014

In some universities, activities such as teaching and so on are explicitly structured into the research program. In others, the formal part of the program consists only of research and the other development activities are something the student needs to independently explore. Either way, these are critical skills that you should acquire before you can embark on a career as an academic. Rather less positively, being a research student is sometimes said to involve 'surviving your thesis', and thus another group of important skills is learning to anticipate and manage stress, and developing the ability to work to long-term deadlines in a sometimes chaotic and high-pressure environment.

These activities—publishing, developing, connecting—are the topic of this chapter.

Disseminating Your Research

The goal of research is to create or modify knowledge. But whose knowledge? It isn't productive to do research and then keep the outcomes to yourself; part of the aim is to make others aware of what you have found. The purpose of research can be viewed as being intended to have *impact*, that is, to change the minds of others. Successful research influences people to behave differently and undertake new activities.

But, you might respond, there are many ways of having influence. In the political sphere, much of what is said is intended to persuade people to have one view or another; the same is true of advertising; and the same is true of all sorts of attention-seeking activities, from fraudsters to alarmists. What makes academic research different is the systems of checks and balances. For example, it is widely regarded as unethical to use the media to publicize research outcomes before the work has been refereed, and there is an expectation that published results are the outcome of an objective analysis that is consistent with the best practice of the rest of the academic community. The influence of work is due to the strength of rational argument that supports it. These kinds of constraints determine how work is disseminated: not in newspapers, or blogs, or mailing lists, but primarily through standard academic forums.

An underlying question is: Why disseminate? There are several good answers to this question.

- To get knowledge of your work into your academic community, as discussed above. I said it earlier, but it is worth saying again: this is why we take the time, not just to write about our research, but to write well. People won't trouble to understand your ideas if they have to struggle with your writing, while clear, lively writing creates the impression that what you say is worth understanding.
- To fulfil the obligation of a publicly funded researcher to make their work widely available.
- To create an academic track record of publications and presentations. Without a track record, it is impossible to pursue an academic career.
- To get feedback on your work as it develops.

Feedback is scary. As researchers, we struggle privately with our ideas, trying to bring them out into the light by writing them down as clearly and articulately as we can. No-one wants to see the outcome of their struggle criticized or ridiculed, and it is only natural that many researchers are reluctant to expose their work in public.[1]

Without feedback, though, we can't learn which of our views is controversial and which 'obvious', or learn how to communicate clearly, or refine our ideas in the light of the perspectives of others. Unless we regard our work as perfect—and I hope it is obvious that no work is perfect!—feedback is essential.

That said, feedback can certainly be unpleasant. After a quarter-century as a publishing academic, ill-thought or aggressive referees' reports can still hurt enough to make me lose sleep. For a junior academic, the wrong kind of feedback can seem crushing, and it is all too common for paper reviews to be intemperate or based on lazy reading of the work. Ultimately, though, you need to remember that, however frustrating it can be to try and get your work in print and get your ideas understood, all good work does get published somewhere—and 'what doesn't kill us makes us stronger'. The right response is to leave your emotion aside (which may take a day or two), and get to work on responding to the feedback as constructively as you can.

Kinds of Dissemination

Researchers use four main mechanisms to tell their colleagues about their work: journal publications, conference presentations, talks in forums such as workshops, and academic seminars.

The different forms of publication are one aspect of academia that really does vary drastically from field to field. In some disciplines, only journal articles are regarded as substantial publications, and conference presentations are little more than an opportunity to talk about current work. In other disciplines, conference papers are seen as at least as important as journal papers, and are much more timely. In some conferences there are fully published, indexed proceedings, and most of the authors get a chance to give a 15 or 30-min talk on their work; in other such conferences, most of the authors present their work as a 'poster', literally by standing in front of a large poster they have designed that summarizes what they have done and explaining it to whoever stops to listen. In such conferences, only a select few are given a speaking opportunity. Historically, in some disciplines published papers

---

[1] Not all researchers are so shy. I had to stop a PhD student in my group, Dave, from posting to international mailing lists to announce his latest discoveries—which were not, on a global scale, all that interesting. I valued his excitement about his work, but his excess of enthusiasm led him to embarrass himself. He needed to develop the patience to follow the ordinary channels of communication. In a similar slip of judgment, Dave decided that a particular group of researchers ought to adopt the referencing style he had learnt during his reading of how-to-write books. He wrote a stinging criticism of some papers, focusing on 'issues' that weren't really problems at all, but just differences in style. Fortunately for him, his criticism was simply ignored.

could not be included in a thesis; happily, I believe this rule is now more or less extinct, although some supervisors would argue that it was a good thing.

Academic seminars, though less formal than journals or conferences, are a vital component of academic communication. Most PhD students are encouraged or required to give seminars in their departments during their studies. I think it is even more important to take the opportunity to give seminars elsewhere, in other universities in your city or places you visit when travelling. I cannot count how often I've heard that a student's work was influenced by comments they got from a group of academics they met while visiting another university. If you make regular presentations, you are likely to sharpen your critical thinking; and the scrutiny your work has undergone in a presentation, both from yourself and from your audience, will bear fruit in your writing. Of course, you need to present to a professional standard, and with confidence.

Some people write books, but this is more typically an activity of an experienced researcher. There is a view in many disciplines that a book should be primarily the product of mature, balanced reflection, not just an opportunity to advance a single point of view. Some PhD students do publish their thesis as a monograph, though, and if you have such an opportunity you should certainly consider taking advantage of it.

A personal perspective: there is no doubt that a long publication list is impressive, even more so when the author is relatively junior. It would be dishonest of me to say that it is clearly in your best interests, some of the time, to publish less rather than more: appointment boards will always be influenced by an appearance of high productivity, and may not really trouble to look at the quality of the publications, especially for a junior appointment. But, as in all things, there is a point at which quantity becomes excess. Fundamentally, to publish you need to have something valuable to say. High-impact publications—those that are cited and remembered—are the product of a considered piece of sustained research, not the outcome of a perceived need to get one more paper in print based on the minimal publishable increment. Again, it would be dishonest to deny that some academics succeed, in part, due to their ability to self-promote; considering how to make others aware of your work is an important part of being an effective researcher. But publication should be a consequence of having achieved good results, not a substitute.

Dissemination Plans

The issue, then, is to choose what to disseminate, and when. If you leave all thoughts of publication until after you graduate, the chances are that you will not publish at all. The way to overcome this is to develop a plan for disseminating material from your project early in your study—perhaps as soon as the topic of your work is clear. The way to think about this plan is as a series of graded challenges, where the aim of each challenge is to capture some key element of the work in the form of a paper or presentation. In my (Gruba's) PhD, the list of challenges I made up as I wrote my thesis had these components:

- Primary research problem.
- Advanced hypothesis and background.
- Pilot study.
- Outcomes of the central study.
- Further directions based on the central study.
- Spin-offs for other areas.

Notice that you could write the material in the first point almost at the beginning of a PhD, and material in the second and third points long before the thesis is finished. Writing this material early will help to shape your thinking.

This is not to say that the structure above will be the same in every discipline—it most certainly won't be. In a technical discipline, it might be that the challenges are an increasingly detailed examination of a single problem; or the outcomes of a series of experiments that offer different forms of evidence about a single hypothesis; or a series of linked pieces of work that together establish a broader result. The point here is that you should be able to formulate some sort of dissemination plan, to be refined as your work progresses, opportunities present themselves, obstacles are encountered, and so on. This planning should closely involve your supervisor; not only will she or he have a broader perspective (think of Dave's naïve behaviour noted earlier), but will be able to introduce you to opportunities you were not aware of, including not just publishing venues but funding for student exchanges and internships.

In your dissemination plan, the first challenge is probably a seminar presentation or short paper that could be written within the first 9 to 12 months of your candidacy. Most departments will require you to hold your first research seminar within this period. Some students develop a couple of variations of a standard presentation about their work—one version for other people in their discipline, another for a more general academic audience. Then, if they get invited to present at short notice, they have something ready. I think the discipline of maintaining such a seminar is a good one that every student should consider.

After your first publication, consider writing another paper every 6 months or so, and integrate the feedback from responses to these papers into your thinking about your project. As you read this you may be thinking, 'but I have a project to do, and very limited time to do it in. I don't have time to waste on writing papers'. Not so: I guarantee that every hour spent on writing papers will make your thesis easier to write, and the act of trying to get a perspective on your own work instead of being continually immersed in it will greatly improve the quality of your thesis.

In the papers that are written early in your PhD, you should focus on only one problem (or theme) at a time, and avoid taking on the 'big picture'; leave that for the last paper you write on your thesis topic. In general, you should prepare such papers with your supervisor as co-author, so discuss your ideas for a paper with her or him, and develop a plan. This will probably consist of developing a list of section headings together, with you writing a draft to the agreed structure. Your supervisor should then criticize the draft as any co-author would, but in addition you can expect to get some guidance about paper writing.

In a paper you are reporting the same material as in a part of your thesis, perhaps part or all of one of your chapters, but to a broader and quite different readership. In the thesis you are addressing the examiners, and your task is to convince them that you know what you are talking about. In the paper you are addressing a far wider range of people. They are reading it because they are interested in your field, and they assume that you *do* know what you are talking about before they even start reading. Indeed, your paper would not have been published had the reviewers not been convinced of this. You are limited to a few thousand words, and you will have to leave out a lot of material that you would include in a thesis.

The challenge then is to tell the story concisely. An introduction to a thesis chapter has the task of telling the reader how the chapter fits into the overall plan, whereas in the paper you are introducing the same material as being important in its own right, so you will have to cover previous work done by others. In the thesis you had written an extensive review of the literature in an earlier chapter. You will have to cover this material in the introduction to the paper, but in perhaps as little as 500 words rather than 10,000. The readers will have to be satisfied with bald statements about this earlier work and your interpretations of it, with deeper critical analysis reserved for a few key points. You then have the challenge of presenting the work, again in a more concise form, and it may be that some lines or argument are only noted rather than explained in detail. The ultimate goal is to produce a piece of work that is reasonably self-contained, with enough narrative and evidence to persuade the reader of its value.

Some conferences publish unrefereed papers, but the principles are much the same. If your work is to be made available to others, it should be cohesive and complete.

Publishing involves choosing a venue, say a particular journal, then shaping your paper to the journal's 'house style'. Most journals have online resources to assist in preparing papers for submission. How to write a paper is beyond the scope of this book, but I do encourage you to read about paper writing, in particular on the challenges of communication within your particular discipline.

Joint Authorship

A challenge faced by research students is of writing papers in collaboration with other people, in particular their supervisors. Writing joint papers is tricky because two or more people are making the decisions. If your joint work is to be really fruitful, you have to acknowledge these difficulties and deal with them. Students need to appreciate that the bulk of the effort may be theirs, while the credit must be shared; supervisors need to acknowledge that, even in the cases where they are largely responsible for the aims and shape of the work, it is nonetheless a shared outcome. Joint publication not only acknowledges the contribution made by your supervisor to the development of your research (and to your development as a research worker), but also commits him or her to a significant contribution to the paper.

Research ethics guidelines have clear protocols for reportage of joint work. These are no more than the commonly accepted rules that permit people to work cordially together, and it is arguably more important to recognize that each of the parties in a joint enterprise will bring to it strengths to address and weaknesses to overcome. In short, a good professional relationship is far more important than a set of rules. Who is to write what? Who will keep the project moving? Whose name will go first? Who will make contact with the editor of the journal? My advice is to resolve these issues openly and early, so that there are no misunderstandings.

Seminar and Conference Presentations

Oral presentations involve several challenges: choosing what to say, transformation of written work into a spoken form, competent delivery, and dealing with nerves.

Your first seminar presentation may be the most difficult one. You are certainly not on top of your project yet, or comfortable with your knowledge of the research area, but you have to convince your university that the project you are working on is suitable for a PhD study, and that you yourself 'have a PhD in you'. How will you proceed?

Don't assume that you can take for granted that your audience will know what the project is all about. Start with your problem statement and aim. Then follow with background, but a much-reduced version of it, because you want to concentrate on your own work. Sketch your ideas and methods, then give a progress report on the results of your own work. In your final seminar before you present your thesis for examination, you should follow the same pattern, starting with the background and aim, but concentrating more on the findings and their implications.

Because one of the intentions of student presentations is to get critical (that is, insightful, honest, and perceptive) feedback, you need to set the tone of your presentation in such a way that you achieve this aim. This means getting the balance right, and not spending too much time on one part to the detriment of others.

In developing a presentation, there are several simple principles to keep in mind. One is that 'a talk is a conversation with educated friends'. You are not giving a political speech, or presenting a legal argument, or convincing people to buy something they don't need or that doesn't work, or trying to crush an opponent in debate, or delivering stand-up comedy, or being a newsreader. That is, there are dozens of kinds of public speaking, and you need to find the right model. In my view, thinking of the presentation as an informal, intelligent explanation is the right one. When you practise your talk, for example, you should be able to model your choice of words on what you would say to your colleagues in the corridor, should they ask you for a quick explanation of what you are doing. There is no need to be excessively formal, or excessively showy.

Another principle is that 'the talk is *about* the work, but is *not* the work itself'. You can't present a complete thesis in 30 min or an hour, so why try? The talk should explain why the work is interesting, what in it is new or insightful, and give the audience reasons to go and read whatever you have written. A talk is a success if it persuades people that this is significant, robust work with worthwhile outcomes, undertaken by a competent researcher; it doesn't need to include every thought or detail. Those are in your writing, and the audience can find them there.

A third principle is 'have mercy on the audience'. You are on a mission to tell a story; make it easy to listen to. This involves considering both the profound and the superficial. *Superficial*: pictures amuse and inform; use them, if it makes sense to do so. Develop the narrative rapidly, and don't spend too long on any one stage of the talk. Engage with the audience and treat them as equals—expect to be treated as an equal in return. Don't read from a script, or worse, from your overheads; there is nothing worse than the presenter who reads from his or her typescript, head down, no eye contact, voice droning, no visual material. Practise, but not to excess. *Profound*: be clear to yourself on what your message is, and design your talk as a logical path to that message. Identify secondary messages, and cut them out unless there is a real need for them. Don't be afraid to expose your uncertainty, or areas where you want advice, or unexpected challenges you have found. Be forthright about the things that didn't work (if they are relevant) or the evidence that pointed the wrong way. Be honest and open.

Gather the material for a talk by brainstorming, just as I described in the context of identifying conclusions in Chap. 9. It helps to be relaxed and not under too much pressure. Your aim is to identify the elements of your work that you particularly want the audience to learn; the talk should be designed around helping the audience to understand these elements and why they are significant or interesting. Simplicity is good; if you think you have too much, you are almost certainly right, and something needs to be removed.

The broad pattern of a talk is much like that of a paper: introduction, background, approach, results, interpretation. But each of these must be approached in a quite different way to that in a paper. The introduction, for example, might begin with a dramatic statement that sets the context for the whole talk. A colleague of mine recently began a talk by saying 'the such-and-such institute just received $10 million in funding for a supercomputer to process this data; with these new methods, maybe one day it could be done on a thousand-dollar laptop'. A strong result allows an opening of this kind; so does a controversial problem. Other topics will have their own strategy, but the fundamental point is that you can design a delivery strategy for a talk on just about anything.

Expect to be nervous. Most speakers are. A key point to remember is that people are at your talk, for the most part, because they expect it to be interesting—they are not there to criticize, or to be aggressive or unhelpful. Likewise, the audience will have plenty of experience of novice speakers, and won't have unrealistic expectations. The best cure for nerves is, one, to know your topic well and, two, to start speaking. If you have brainstormed the talk well, and have good material to speak to—and, ahead of time, you should have ensured that you do have something sensible to say about every slide—then those nerves should quickly ebb away.

As mentioned above, you should never read your notes aloud, or read from your slides. By all means use notes to prompt yourself (though many people find them more of a distraction than a help), but direct reading rarely works, even if you have used a professional scriptwriter. And, while I am on this topic, when you are speaking *never* turn your back on the audience. Don't hide from them; face them, and make them want to listen to you.

Being a Graduate Student

The primary reason to undertake a graduate degree, particularly a PhD, is to establish yourself as an effective researcher, but a broader outcome is that a PhD prepares you for life as an academic. To a lesser extent this is also true of Masters degrees and minor research theses: completion of a thesis hones your ability to make judgments and work independently, and shows that you have skills that are essential to success as a researcher.

But being a researcher involves other skills too. Some of these are close to the core activity of undertaking research, such as refereeing of papers that other researchers have submitted to conferences or journals. Effective refereeing (or reviewing) is a genuine challenge. Ideally a referee would be expert in every aspect of the paper being examined, but in practice this is rarely the case, and often there is no such expert! Thus the referee can be in the awkward position of having to make an informed judgment on a paper, while, in all likelihood, knowing less than the authors about some aspects of the research area. This judgment needs to identify the flaws that might prevent publication while ensuring that genuine innovation is recognized rather than ignored due to trivial failings. Another intimidating aspect of refereeing is that it can involve making a decision about the quality of work of people who are considerably more experienced than you are.

However, writing of referee reports is excellent training for the task of writing your thesis's literature review. You may get to see the reports written by the other referees, which is usually an interesting experience if only because of the extent to which referees notice different problems and form contrasting opinions. If the article is revised, and sent back to the referees for further review, you will have an opportunity to see how your criticisms are received and responded to—sometimes these responses will be reasonable, and sometimes they won't.

Your lack of experience may make the task of reviewing more difficult, but it should not stop you taking on refereeing assignments. So long as you are honest about the extent to which you are knowledgeable—and, for example, any diligent reader can make useful comments about readability or the completeness of a bibliography—the editor will appreciate your efforts.

Teaching is a key academic skill. Acquisition of teaching experience is not a key part of doing research, but learning to communicate is—and if you plan to continue in academia, it may be essential to have a track record of teaching. Delivery of lectures or tutorials, development of teaching materials, and related activities such as one-on-one coaching, are all effective ways of getting feedback on your communication skills. One thing that I have noticed is that teaching is a great way of building your confidence, not just in public speaking but in general interaction with other students and academics. If it makes sense to take on teaching assignments during your research then my suggestion is that you do so.

Another academic activity is mentoring. It may not at first be obvious that learning to mentor is an important part of being a research student, but I think it is a core skill. Why? One reason is that a large part of succeeding as a research student is your interaction with your supervisor—whose guidance is a form of mentoring. A wonderful piece of advice I was given early in my PhD was to learn to ask the questions my supervisor would ask, to anticipate what my supervisor would want, and to solve issues that my supervisor would be concerned about. In other words, I was being advised to try and put myself in my supervisor's role, and regard him as leading by example. From this perspective, research study is a form of apprenticeship, where skills are passed by practice and example from master to novice. By becoming an academic, you must also become a mentor.

Opportunities for mentoring vary from discipline to discipline, and might include coaching of undergraduates, involvement in small research projects, or partnering with junior research students. Such mentoring can have many benefits, not least of which is that they can lead to lifelong working relationships.

As a supervisor I notice that my students who are themselves mentors are better than other students at understanding what I require of them. That is, their experience of mentoring may be helping them to understand the student–supervisor relationship from both perspectives, and to build a more effective partnership with me.

Such students are good at knowing when to seek advice—and will happily seek it when appropriate to do so—and are also good at knowing when guidance should *not* be sought, that is, they should try to answer their questions for themselves. It is not clever to batter away at a problem if a few minutes of someone's time is all that is needed to point you in the right direction, but nor is it clever to ask for guidance on every little momentary issue that troubles you. A mature student sets problems aside for later consideration, resolving some and sharpening others for future conversations with a mentor, rather than seeing every unknown as an obstacle for which guidance is required.

Effective Research

What are the skills of an effective researcher, and how are they acquired? The answer to this question lies in the background of the students who enter research degrees, which, quite simply, is highly variable. People who decide to undertake research may have come straight from another degree (which might have been their first degree after secondary school), or may have been in the workforce for decades in one capacity or another. Thus a new student might be expert in recent academic knowledge, but inexperienced in terms of independence and skills such as writing; or may be skilled in the practice of their discipline, but out of touch with the latest developments. The research may be interdisciplinary in some way (for example, technological research projects with applications in medicine), and thus the student may be highly knowledgeable in aspects of the work but have no background in other aspects. Some PhD students have experience of research, which however may be limited to a few months of closely supervised work in a tightly defined project; others have broader experience, but have never grappled with the difficulties of undertaking a project independently.

Thus, in the course of the research degree, not only must you undertake research, but consciously seek to identify where you are weak and design ways of becoming more competent. You might for example take an undergraduate subject, set yourself a program of reading in an unfamiliar area, or establish or join a study group. At the start of your project, as well as developing research questions and getting familiar with the literature, you may consciously decide to learn basic skills, by for example working through elementary tasks in the lab, exploring a document archive, or establishing an effective online working environment.

A part of the learning during research study flows from the kind of activity and thinking that the doing of research entails. It is an essential part of research that it involves steps into the unknown, and such steps are often mistakes. Learning to deal with mistakes, which may have taken months to discover and resolve, is part of the process. Another part of learning is taking of new responsibilities—specifying the tasks as well as undertaking them, and working with less hands-on supervision. Then of course there are all the technicalities of working within your discipline, including lab work, investigation of primary sources, development of designs and studies, and so on. In some PhDs, there is a cyclic pattern of starting new investigations under increasingly light supervision. By the end, a strong student may well be initiating, designing, evaluating, and reporting a complete investigation, with the supervisor in the background in a relatively passive role.

I feel as if I have been a little bit sneaky here, because hopefully you have been led to a new way of thinking about what a PhD or other research degree is. Students enter a PhD wanting to 'do research' or gain a qualification, but in addition a research degree does a great deal more. Arguably, the most important outcome is that a PhD creates a maturing, independent researcher, one with a sense of what problems to work on; with good knowledge of their own limitations; with experience of clear thinking and rigorous argument, systematic organization of ideas, critical analysis, and communication; who has the skill of separating passion for their work from objective assessment of its value; and who is ready to lead the research and to design a process that is intended to settle a question.

The Arc of a Research Degree

A minor thesis may only require a single semester, and completing it is a sprint from start to finish. Three or four months is not a long time to find and read the key papers in an area, undertake a study, and produce a polished report on the work. Most students find that they have to be highly focused from the first week, with a systematic working schedule that fills not just their days but to some extent their weekends and evenings. Such a working pattern can be stressful, but is quickly over.

For longer research degrees, in particular Masters and PhDs, a sprint cannot be successful; few students have the energy or resources to work 70 or 80 h a week for a year or more. Such degrees truly are a marathon. An effective pattern is as follows. At the beginning, expect to work at a steady pace—if you are studying full-time, this will be little different, in most cases, from the kind of pattern people have in an office job. Maintain some outside interests and a social life, but make sure they do leave enough time for your study.

For research students who have come straight from an undergraduate degree, such working habits may be a big change. With few deadlines or regular commitments such as lectures, the pressures that drove engagement with study are completely removed. A student on a scholarship, who may have had irregular part-time work, may also be free of the need to follow a regular daily routine, and worse, very possibly developed a habit of working a light week during semester and cramming during exams.

My student Jack came to his PhD with bad habits of this kind, and in particular had developed an extraordinary number of activities outside study, including competitive sport (inline skating), performing in the local music scene, writing educational software, and a full list of social engagements with his soon-to-be wife. He was also something of an addict of computer gaming; coping with this problem is a real challenge for any student whose study involves sitting in front of a computer—rather like being an alcoholic who has to work in a bar. Somehow he had maintained all of these things throughout his undergraduate years, in part, I suspect, because the stop–start pattern of coursework meant that he could switch between his different interests. As a research student, it soon put him on a course for failure; he would have gotten into difficulty with far less distraction that this. He eventually resolved his issues by taking a year off to build his bank balance; taking his marriage seriously (he adopted a more responsible attitude because of the sacrifices his wife was making for his studies); giving up the software development; and scaling back the music to a couple of gigs a month. The most important change was a resolve to be in the lab or at his desk for eight hours a day, usually from nine to five, with other time completely free to use as he chose. He finished with a strong thesis, but, including the year off, ended up taking 2 years longer than he had originally planned.

The first months of a Masters or PhD tend to be fairly exploratory, not just of the topic but of what works and what doesn't in terms of working habits. I discussed some of these issues in Chap. 4. Part of the solution is to begin to build a realistic schedule of work, which should become increasingly detailed as you progress. Tasks for the next few months should have weekly or fortnightly milestones; if they don't, they need to be further broken down. This schedule should evolve into a timetable for submission, focused on your planned completion date. (And it should include some planned breaks from study of a week or two. We all need holidays). Use this initial period to develop working habits, and life habits, which will take you through not just the early familiarization phase of the degree but also the 'long middle', that period that seems distant from the start and distant from the end—when the work has become so routine and familiar that some students despair of anyone ever finding it interesting. But do not worry, as you write-up the significance of what you have done will soon become obvious.

The final phase, perhaps as long as 9 months in a PhD, is very different and requires total commitment and focus. As you create a complete thesis you need to be familiar with every detail of your work as presented in a sequence of chapters, and this requires removal of distractions. You may have to give up your outside interests and social life, and make sure that your partner, or children, or parents, or whoever it is you live with, is aware of your needs and constraints. This phase can be very stressful—I've known students who share an office to come to blows over trivialities such as a habit of tapping a pencil on a desk—which means that you need to learn to identify safety valves and when to make use of them.

This may sound surprising, but you may also need to prepare for life after submission. If you have worked all your waking hours for months on end to get a thesis finished, submitting it can leave a void in your life. Line up activities such as part-time work, or plan a holiday. Even a small break can make a big difference.

Summary of Chapter 12: Beyond the Thesis

Publishing:

* Make sure that the good work you are doing in your project gets published.
* The only way to ensure this is to have a dissemination plan. The plan should be geared to publishing while you are still working on your project.
* Regularly give research seminars. Doing so gives you feedback on your project, and will also form the basis for conference papers and papers in learned journals.
* Research papers have a different set of rules and conventions from theses. To publish papers, you need to address a different readership using these different conventions.
* Papers written on the basis of work done as part of your project should in general be written jointly with your supervisor and possibly other colleagues too. You need to do this in a way that respects the input of both parties.

Spoken presentations:

* Spoken presentations are entirely different from written presentations of the same material. Never just read your written paper to the audience, or even the same paper with bits left out.
* Take the effort to develop your skills as a speaker; these skills will be essential to your professional life.

Being a graduate student:

* Think and act like the professional researcher that you are striving to become, taking responsibilities for your work, seeking collegial advice when needed, and maintaining a sustainable regimen of work. Seek to rectify your weaknesses.

- Consider taking on other academic commitments, such as teaching, reviewing, and mentoring.
- Balance life and work commitments, paying particular attention to the demands of partners and other family.
- Anticipate that 'life' will seemingly get in the way at times, and learn to cope.
- Expect the write-up phase to be a committed, focused slog where you have eliminated distractions.

Online resources. There are many excellent websites on how to give spoken academic presentations, including material on topics such as:

- Overall techniques for giving talks.
- Learning to speak with confidence, and overcoming stage fright.
- Effective data presentation.
- Good style in slide design (regardless of whether your preferred tool is Power-Point, OpenOffice, Beamer, or whatever).

Appendix

Afterword

The creation of *How to Write a Better Thesis* has, in part, been an outgrowth of my teaching of research methods. For a long time my approach was to present it as a series of tasks such as starting the writing process, designing experiments, and interpreting results. But as a supervisor, and a participant in a variety of student progress committees, I had found myself asking questions about the students I encountered, about why they did or didn't do well in their studies. I noticed that their strengths and weaknesses—the things that led to success or failure—often fell outside the topics I was teaching as research methods.

For example, one question I asked was *What were the qualities that I admired in the most successful students?* Brilliance, sometimes, but more often an ability to 'get the job done'; that is, the quality of battering at problem until it was defeated. Brilliant students may get there faster, but a determined, careful student will find a range of ways to approach a problem and methodically work through them, often to equally good results. *What was it that led me to judge them 'successful'?* Strong or numerous publications, sometimes, but more often perseverance in the face of difficulty, a determination to work around their own shortcomings. That is, often these students were not at first glance the most talented, but had a spirit that made them able to do first-class research despite gaps in their abilities. Which leads to *What really are the key skills of a successful student?* Ability, sometimes, but more often a combination of hard work, exploratory thinking, and a willingness to learn from mistakes.

Other questions were less positive, but equally pressing. *Why do good students sometimes fail? Why do some student–supervisor relationships break down? Why is it that the experience of being a research student is so stressful for some people?* It seemed to me that such questions, in contrast to questions about the mechanical steps of getting it done, were not discussed often enough. This book is, in part, an attempt to answer these questions, and to help students succeed in the face of difficulties and challenges.

How to Write a Better Thesis is also a response to my experiences as a supervisor. Looking back at the many PhD students I have supervised or worked with, there are several aspects that are striking. One is the diversity of experience and background that

D. Evans et al., *How to Write a Better Thesis,* DOI 10.1007/978-3-319-04286-2,
© Springer International Publishing Switzerland 2014

they brought to their study, from undergraduate education at top-ranked universities in some cases, at rural technical colleges in others; from no experience to 25 years in the workplace; from stable family life and a standard school–university–research transition to teen years spent in blue-collar work (without finishing high school) followed by workplace experience and graduate certificates; from speaking English as a first language with long exposure to academic writing to learning English as they did their research. Another is the extent to which any of these students could succeed regardless of background; while an 'easy run' can be helpful in some respects, a student who has struggled to have the opportunity of undertaking research has probably learnt skills—such as a particular kind of persistence—that more than compensate for other disadvantages. A third aspect is the way in which their skills tend to converge during their study, as they make use of their strengths and rectify their weaknesses. Best of all has been the fact that the majority of them have gone on to do research, either in academia and industry, and have continued to develop. I'm still in touch with most of them.

Perhaps two or three of these students, but no more, were adequate as academic communicators before they began their PhD. I'd say that much the same is true of the research students who have gone through my department; considering the couple of hundred who I've taught in research methods, no more than ten or so struck me initially as already good at writing and presenting. Yet the great majority of them acquire the skill, and confidence, to write well by the time they submit. (Acquiring the skill of spoken presentations seems to take longer, but during their PhD most students certainly improve.) Indeed, to go further, I would say that *every* student who makes a serious effort to learn to write largely succeeds in doing so.

There's no doubt that the task of assembling and finishing a thesis is a transformative one. At the time of submission of my own thesis, I (Zobel) felt that it was a success against the odds. I came to research with some years of experience in the workplace and an over-confident belief in my ability to communicate. Under the guidance of my supervisor, I was introduced to the challenges and methods of technical writing, and gradually realized that, in this domain, I was not a good communicator at all. I had to swallow my pride, go back to basics, and struggle to develop an effective approach to getting a thesis written. At times I thought I would not succeed, hence my relief when I finally submitted.

Now, though, with the wisdom of hindsight, I don't think that my struggle was so special. Most of the students I've worked with seem to go through something similar. Maybe their insights and paths to success are different to mine—in my case, it was the experience of feeling uneducated on the topic of communication that led me to start teaching others about writing and research methods—but the differences are less obvious than the similarities.

A painful bit of hindsight is that my PhD thesis is—how can I put this nicely?—not an excellent piece of writing. I developed while creating it, but have developed far more since then. That is, the process of becoming a better author and researcher is ongoing, so on the one hand I sometimes cringe when I read older papers of mine and on the other am grateful for the ease with which, on a good day, I can get some writing done. Hopefully this book has helped you along a similar path.

Appendix: Analysis of Examiners' Reports

John McDonald of the University of Ballarat undertook an analysis of examiners' reports submitted to his university, and identified numerous common elements that for the examiners were characteristics of poor or of strong theses. A digest of these is listed below (used with permission; thanks John). These points are at a mix of levels of significance and breadth, but they are all valuable. Think of them as a checklist.

A key message that is worth highlighting is the extent to which examiners felt that the ultimate quality of a thesis is largely determined in its formative stages—I agree! A great result requires that you make a good start.

Characteristics of a High Quality Thesis

- The title clearly reflects the focus and the argument.
- A significant and substantial problem has been selected for investigation.
- There is an early statement of the project aims.
- The project presents a considerable advance on existing knowledge.
- The thesis demonstrates a systematic pursuit of a consistent line of inquiry.
- It is well-planned and executed, with each section clearly building on the last (that is, there is a coherent and unifying macro-level structure).
- There is clear signposting and linking between paragraphs, sections, and chapters. It consistently (but not repetitively) reminds the reader of the purpose, argument, or overall thrust of the thesis.
- The literature review is critical and evaluative, and sets forth an argument for why and how the study should be conducted.
- The discussion of the rationale for selecting a methodology and method (including up-to-date methodological literature) is balanced. The ground-setting is sophisticated and appropriate (including exposition of underlying assumptions, and relevance to the research aim).
- The research design is appropriate and allows the questions to be answered.
- There is a meticulous account of the procedure.
- A rich variety of evidence is employed to develop a balanced argument.
- Advanced analytical skills are used to demonstrate a deep understanding of the problem; a clear chain of evidence is laid down.
- The discussion is disciplined and not excessively speculative.
- Conclusions are well drawn and convincing (they relate the outcomes back to the research aims); clear and strong knowledge claims are made about the exact contributions of the thesis.
- Key concepts or variables are clearly defined and consistently used throughout.
- Written expression is elegant, precise, and economical.
- There is evidence of systematic proofreading and error correction.

To these, I would add that it is particularly impressive to receive a thesis that is the product of thorough work, in the sense that discussions are considered and insightful rather than superficial, and key arguments have been diligently explored. I try to not be too critical of presentation (in particular because the majority of theses I have examined are by students whose first language is not English), but I do value a thesis where the copy-editing is careful and significant effort has gone into creation of figures and tables that are easy to understand.

Characteristics of a Poor Thesis

- Objectives and protocol of the study are not stated.
- The research questions are either not significant or are self-evident (no risk of a successful outcome).
- The principal purpose or argument of the thesis is difficult to discern.
- No clear delimitations to the study.
- Overly simplistic comments and generalizations.
- The scope of the thesis is overly ambitious.
- Grasp of the literature has serious limitations (the student is unaware of major relevant works, or uses older works that are no longer authoritative or never were authoritative).
- The description of the literature is serial rather than interpretative (with scant critical analysis or argument emerging).
- There is no clear connection between the focus of the study and the logic or foundations of the research on which it is based.
- Theoretical perspectives or conceptual frameworks are left implicit; the rationale for a particular theoretical approach is missing or undeveloped.
- Shows no awareness of the alignment or compatibilities of particular theoretical and methodological approaches.
- The overview of theory is broad and lacks depth or persuasiveness (especially noted by a reliance on undergraduate texts without reference to primary authors).
- The description of the sample selection strategy is inadequate (inclusion and exclusion criteria not stated).
- The arguments are intrinsically weak.
- Large slabs of (qualitative) data are used to present a point when smaller excerpts with richer or deeper analyses are needed.
- No demonstrated understanding of appropriate statistical analyses and interpretation, or insufficient detail on how the data analysis was undertaken.
- Triangulation often claimed but rarely delivered.
- Contains sweeping, unfounded conclusions that have little or no basis in evidence.
- Definitions of key terms are either omitted or imprecise.
- Contains poor photos, confusing diagrams, and inadequately labelled tables.
- Contains poor written expression that detracts from the candidate's argument. Littered with spelling and typographical errors; has incorrect or inconsistent referencing.

- The text is unnecessarily long and wordy. Material is repeated.
- Lack of critical self-evaluation of the research.

There are several aspects of poor theses that I find plainly bewildering, but that do seem to be common. In addition to the issues listed above, I note: descriptions of processes that cannot be understood; theses that seem incomplete, with some entire component missing (most damning is a lack of critical analysis of the work presented in the thesis, or even a complete absence of discussion of results); insufficient data to support the conclusions, or indeed any concrete conclusions at all; whole bodies of work unreferenced, despite obvious relevance; and persistent 'microgarbling', in which sections and even paragraphs don't have a clear thread of ideas, but instead are just a jumble.

I suspect that many such theses are a consequence of the student simply having run out of time. If there is one single lesson I have learnt from examination, is that starting the thesis early is not just important, but is critical. If you are doing a research degree and haven't yet begun to write your thesis, don't delay any further!

Notes on Further Resources

Since the publication of the earlier editions of *How to Write a Better Thesis*, the web has become a primary tool for finding and distributing scholarly information. Most researchers are aware of the web as a source of knowledge in their discipline, but it is also an excellent source of knowledge about being a researcher. There are a great many online resources, including lists of texts on dissertations and postgraduate study, blogs on scholarly writing, and numerous guides maintained by organisations such as university libraries and research offices. I suggest searching with terms such as 'dissertation writing', 'surviving a thesis', 'how to write a thesis', 'scholarly writing', 'academic presentations', or 'presentation skills'. You should also search for guidance related to your specific discipline or approach; examples include 'social science research', 'qualitative approach', or 'health science research methods'.

Despite the growth of the web, however, books continue to be published—including this one!—and for good reasons. A well-designed book provides a consistent, authoritative, and thoughtful point of reference, in ways that a dynamic, fluid web resource cannot. Again, web search is an effective way of finding such books, which range from general advice to discipline-specific texts. Some disciplines have long-standing, comprehensive style guides; if there is such a guide in your academic area, you should make use of it. I also suggest that you find a good book on the mechanics of writing, and another on writing style. These skills complement the approach I have taken here, of helping students to succeed through discussing the challenges of the task of writing a thesis.

The following are examples of books that I feel are of enduring value—as you will notice, several of them have been through multiple editions. I have used most of these over many years. (If you search for online guides where these are recommended, you will quickly discover other texts on similar topics.)

Booth, W, Colomb, G and Williams, J, *The Craft of Research*, 3rd edn, Chicago University Press, Chicago, IL, 2008.

Day, A, *How to Get Research Published in Journals*, 2nd edn, Gower, Aldershot, UK, 2007.

Day, R, and Gastel, B, *How to Write and Publish a Scientific Paper*, 6th edn, Greenwood Press, Westport, CT, 2006.

D. Evans et al., *How to Write a Better Thesis,* DOI 10.1007/978-3-319-04286-2,
© Springer International Publishing Switzerland 2014

Phillips, EM and Pugh, DS, *How to Get a PhD: A Handbook for Students and their Supervisors*, 5th edn, Open University Press, Buckinghamshire, UK, 2010.

Rudestam, KE and Newton, RR, *Surviving Your Dissertation: A Comprehensive Guide to Content and Process,* 3rd edn, Sage, Thousand Oaks, CA, 2007.

Rugg, G and Petre, M, *The Unwritten Rules of PhD Research*, Open University Press, Berkshire, UK, 2004.

Silvia, PJ, *How to Write a Lot*, American Psychological Association, Washington, DC, 2007.

There are of course many further books on these topics, some excellent, some not. As a very general piece of guidance, I find that texts that advocate a 'recipe' approach to thesis writing ('precisely follow these simple instructions and you cannot fail') are less successful than those that are more collegial and advisory in their approach. They are also less successful than those whose aim is to simply explain the components of a thesis, and leave the topic of the challenges of writing to texts such as this one. I have also used a wide range of style guides and so on, which in my view are best explored through a visit to your local library or bookshop; thus I do not list them here.

Another place to look for advice is in journals such as *Studies in Higher Education*, which has published papers such as Mullins, G and Kiley, M, '"It's a PhD, not a Nobel Prize": How Experienced Examiners Assess Research Theses', vol. 27, no. 4, 2002, pp. 369–86, and Holbrook, A, Bourke, S, Fairbairn, H, and Lovat, T, 'Examiner Comment on the Literature Review in PhD theses', vol. 32, no. 3, 2007, pp. 337–56. Other journals have similar scope. These are readily found with the usual resources, such as the scholar-specific tools provided by web search engines or the search tools at a typical university library.

To become strong at scholarly communication, you need to read widely. This includes not just dissertations and papers in your field, but a range of perspectives on topics such as strategies for research and communication skills. No one book, not even this one, is sufficient by itself. I continue to develop my own skills through such reading, and I encourage you to do the same.

Index

D. Evans et al., *How to Write a Better Thesis,* DOI 10.1007/978-3-319-04286-2,
© Springer International Publishing Switzerland 2014